D1221190

Framing Production

Inside Technology

edited by Wiebe E. Bijker, W. Bernard Carlson, and Trevor Pinch

Janet Abbate, *Inventing the Internet*

Charles Bazerman, *The Languages of Edison's Light*

Marc Berg, *Rationalizing Medical Work: Decision-Support Techniques and Medical Practices*

Wiebe E. Bijker, *Of Bicycles, Bakelites, and Bulbs: Toward a Theory of Sociotechnical Change*

Wiebe E. Bijker and John Law, editors, *Shaping Technology/Building Society: Studies in Sociotechnical Change*

Stuart S. Blume, *Insight and Industry: On the Dynamics of Technological Change in Medicine*

Geoffrey C. Bowker, *Science on the Run: Information Management and Industrial Geophysics at Schlumberger, 1920–1940*

Geoffrey C. Bowker and Susan Leigh Star, *Sorting Things Out: Classification and Its Consequences*

Louis L. Bucciarelli, *Designing Engineers*

H. M. Collins, *Artificial Experts: Social Knowledge and Intelligent Machines*

Paul N. Edwards, *The Closed World: Computers and the Politics of Discourse in Cold War America*

Herbert Gottweis, *Governing Molecules: The Discursive Politics of Genetic Engineering in Europe and the United States*

Gabrielle Hecht, *The Radiance of France: Nuclear Power and National Identity after World War II*

Kathryn Henderson, *On Line and on Paper: Visual Representations, Visual Culture, and Computer Graphics in Design Engineering*

Eda Kranakis, *Constructing a Bridge: An Exploration of Engineering Culture, Design, and Research in Nineteenth-Century France and America*

Pamela E. Mack, *Viewing the Earth: The Social Construction of the Landsat Satellite System*

Donald MacKenzie, *Inventing Accuracy: A Historical Sociology of Nuclear Missile Guidance*

Donald MacKenzie, *Knowing Machines: Essays on Technical Change*

Donald MacKenzie, *Mechanizing Proof: Computing, Risk, and Trust*

Maggie Mort, *Building the Trident Network: A Study of the Enrollment of People, Knowledge, and Machines*

Paul Rosen, *Framing Production: Technology, Culture, and Change in the British Bicycle Industry*

Susanne K. Schmidt and Raymund Werle, *Coordinating Technology: Studies in the International Standardization of Telecommunications*

Framing Production

Technology, Culture, and Change in the British Bicycle Industry

Paul Rosen

The MIT Press
Cambridge, Massachusetts
London, England

Set in New Baskerville by The MIT Press.
Printed and bound in the United States of America.

Library of Congress Cataloging-in-Publication Data

Rosen, Paul, 1963–
Framing production : technology, culture, and change in the British bicycle industry / Paul Rosen.
p. cm. — (Inside technology)
Includes bibliographical references and index.
ISBN 0-262-18225-4 (hc. : alk. paper)
1. Bicycle industry—Great Britain. 2. Bicycle industry—Technological innovation—Great Britain. I. Title. II. Series.
HD9993.B543 G77 2002
338.4'76292272'0941—dc21

2001056240

for Grandma and Grandpa, and their house full of books

Contents

Acknowledgements

This book began life as my Ph.D. thesis, which I researched and wrote while at the School of Independent Studies and the Centre for Science Studies and Science Policy at Lancaster University between 1991 and 1995. My studentship was funded by the joint ESRC/SERC committee for Science, Technology and Society studies (SERC award number 91302983). I'd like to thank my supervisors Mike Michael and Brian Wynne for help and advice and for asking difficult questions throughout that process, and the postgraduate community at CSSSP for camaraderie, for sharing ideas, and for generally combining academic support with friendship. I'd also like to thank my examiners, John Law and Karen Legge, for a gentle *viva voce* and for stimulating my thoughts about publication.

The process of turning the thesis into a book has spanned much of my time as a research fellow at the Science and Technology Studies Unit, first at Anglia Polytechnic University and now at the University of York. Grateful thanks are due, therefore, to my colleagues at SATSU for encouraging this, for reading various drafts of book chapters and related conference papers, journal papers, and book articles, and for not becoming as skeptical as I have sometimes been that I'd ever finish the thing. Andrew Webster deserves special thanks for all of these things and for his willingness to give time to supporting my work while also managing numerous other projects in a busy and thriving research unit. Other colleagues too have commented on various papers and chapters over the last few years—Annemiek Nelis, David Skinner, Janice McLaughlin, Nik Brown, and especially Brian Rappert. Brian had the misfortune to share an office with me for much of the time I was writing and revising the book, and he was also kind enough to read the previous draft in its entirety. I haven't followed all (or many?) of his suggestions as to how to revise it, but could not have reached this point without his help. In addition, the book has benefited greatly from my work with Andrew, Janice, and David on our "technology acquisition" project, especially in improving my understanding of organizational change. Andrew's and David's support and help in developing new avenues of research in the areas of transportation and land-use planning and policy has also helped me turn the vague conclusions of my thesis into the better-thought-out discussion that makes up chapter 7.

It seems fitting that as I was completing this manuscript SATSU was relocating to the University of York, where as a Master's student I first encountered the sociology of technology and theories about modernity and postmodernity. I'd like to thank Trevor Pinch (who taught in that Master's program) and Wiebe Bijker, who both deserve credit for providing the stimulus that set me off researching the bicycle industry and for their encouragement and support in many ways as the project progressed. Wiebe was instrumental in getting my book proposal accepted and in encouraging me in the rewriting process. I am also grateful to Colin Divall for providing valuable comments on my original manuscript.

Numerous others with whom I have shared thoughts and ideas at STS conferences have also contributed in various ways to the final product. This is especially so in the cases of the Prometheus Wired conference (Munich, October 1998), the International Summer Academy on Technology and Sustainability (Graz, July 1999), and a visit to the Centre for Studies of Science, Technology and Society at the University of Twente in September 1999. I thank Peter Lyth, Helmut Trischler, Harald Rohracher, and Adri de la Bruhèze for organizing these events, where I presented work that helped take my thinking for the book beyond the boundaries of the original Ph.D. thesis and where I received questions and comments that challenged my ideas in ways that were at once provocative and dispiriting.

Beyond the STS community, I was surprised and pleased while carrying out my research to find myself a member of a small community of bicycle researchers, both academic and amateur. Thanks to Nicholas Oddy for first making this connection for me, and to Andrew Millward for sharing research data, especially regarding the early days of Raleigh. The same—and more—goes for David Patton, with whom I have had numerous discussions in Cambridge, in Reno, and via transatlantic emails about bicycles, STS, and academic and non-academic life. Rita Prestwich deserves credit for making the connection between mountain bikes and postmodernity before I did. I'd also like to thank members of the Veteran-Cycle Club—notably Chris Watts, Mac Mumford, Nick Clayton,

John Skeavington, and John Malseed—who generously replied to my requests for information in their newsletter when I came to dead ends in my research.

A good deal of my field work on the cycle industry was based on archival sources held at the Nottinghamshire Archives and at the Nottingham Local Studies Library. My grateful thanks go to the staff at both places for collecting endless documents and answering my questions without complaining, and to Lisa Warburton at the Cyclists' Touring Club for giving me access to the club's archives. Thanks also to Carlton Reid of *Bicycle Business* for inside information about the industry, especially his careful charting of the recent troubles at Sturmey-Archer and Raleigh.

I could not have completed the project without the cooperation of people in and around the bicycle industry who gave me their time in interviews or conversations, or by sharing information with me in other ways. These people are identified where possible in appendix A, but special thanks to all the staff at Raleigh who gave me their time in numerous interviews and factory tours. I'd also like to pay tribute to the work and commitment of Kath Hamer of York Cycleworks, who was killed in a car crash not long after I interviewed her.

Grateful thanks to Raleigh Industries, Nottinghamshire Archives, Gary Fisher, Specialized (UK), Rob Van der Plas, and Bicycle Books for permission to reproduce copyright material.

My final acknowledgement is to my wife, Buffy Breakwell, to whom I am eternally grateful for sharing her life with me (and for turning my life upside down in the process), for introducing me to the sociotechnology of transportation in Thailand (where some of the later chapters in the book were written), and for her willingness to occasionally go cycling with me against her better judgement. I get a kick out of you too, Buff ;)

Where British and American English differ, American usage (e.g., 'transportation', 'automobile') is generally favored, The MIT Press being an American house. In the text, dates are given in the American style; in citations of archival materials, the British style is used in cases where that style is used by the archive.

Framing Production

1

Technology, Culture, and the Politics of the Bicycle

The Personal, Political, and Intellectual Contexts of the Bicycle

Bicycles have many meanings. They were for many of us a treasured part of childhood, marking the stages of our growing independence—our first experience of being beyond immediate parental control in the park or local streets (even if our parents were only a few hundred yards away); the fear and excitement of achieving independent balance when the training wheels were first removed; traveling to school or elsewhere by ourselves or with friends, no longer dependent on the good will and driving skills of adults; and perhaps the means to earn some money of our own delivering newspapers. Unfortunately, in recent decades such experiences have become a source of worry for parents. Perhaps bicycles are no longer so central to childhood independence as they once were. Adult cyclists have become increasingly rare. Yet not so many years ago, cycling was the main form of utility transportation for working people, and a major leisure activity too.

Nevertheless, new meanings associated with health and fitness, leisure, and the environment have begun to spring up. Cycling, which once symbolized children's independence, might now mark a shared family experience, such as Sunday rides away from the traffic, perhaps on one of the many leisure cycling routes that have been built recently by groups such as the UK charity Sustrans. Cycling has also been appropriated by the "green transportation" lobby as a solution to traffic congestion and pollution. In many cities it will get you to work more quickly than other transportation modes, and more healthily: although cyclists are vulnerable to exhaust fumes, these are worse for those sitting in the closed environment of a car, and the cyclist has the added bonus of getting exercise without having to go to a gym.

Alongside these newer developments in the meaning of bicycles are longer-established and more specialized ones. Engineering enthusiasts have been developing alternative designs for bicycles and their components and accessories since the bicycle's earliest days. Sometimes they do this for the sake of technical improvement—for greater speed, less wind resistance, or greater comfort or safety; sometimes they do it simply for fun. Occasionally they do it with a social goal in mind—to remove the moral or material obstacles facing women riders, to help give some independence to the disabled, or just to provide an alternative to the private automobile. And hobbyists and amateur historians have been charting (and arguing about) cycle history and collecting vintage machines for almost as long. In recent decades their numbers have grown substantially, accompanied by exhibitions, displays, conferences, and vintage cycle rides (complete with authentic costumes).

Cycle sport may be the aspect of cycling most visible to the general public. Established races such as the Tour de France and Olympic pursuit cycling and newer competitions for BMX and mountain bikes have together seen a growth of interest, especially since the 1980s. This has been fueled both by the positive environmental image of cycling and by the rapid technological change that makes every race or exhibition a spectacle featuring the latest innovations, such as the Lotus Sport "Superbike" on which Chris Boardman won the 1992 Olympic pursuit competition. All this is helped by the specialist and the popular media and by the sports promotion industry, which together further encourage the development of highly specialized niche markets and audiences.

Personal engagement with technology in hobby, sport, or professional activities can be crucial in giving it meaning (Pacey 1999). There is a great deal of literature on the various meanings of bicycles—general accounts of the technical development of the bicycle or the social history of cycling; histories of significant firms, machines, and sporting events; biographies of prominent cyclists, designers, and entrepreneurs; cycling novels; health and transportation policy reports; glossy coffee-table books; and manuals for building and repairing bikes and their components.[1] However, this book is concerned primarily with bicycle *production*, which has been somewhat neglected in the cycling literature. Bicycles are, like other consumer objects, products of an industry based on complex networks of supply and demand—an industry dependent on fickle consumers who need to be cajoled into changing brand loyalty or upgrading equipment that could easily last a few more years. Innovation and change are thus as central to bicycle production as they are to cycle sport. Indeed, innovations tested on the racetrack often are adapted for the consumer market.

Product innovation in the bicycle industry has, throughout its history, been part of a much wider process of change—a process that combines new product designs with new materials and production methods, new ways of organizing firms, new ways of addressing consumer demand, and new social structures and cultural values within and beyond the bicycle market. This book is, then, not only about bicycles and cycling; it is also about the interaction of technological, industrial, organizational, social, and cultural change.

The book focuses mainly on two periods of change in the British bicycle industry—periods in which multiple factors (the modernization of production and of the ways in which the industry organizes itself, changing relations among industrial players, the link between labor relations and production innovations, the interaction between changing markets and product development) came to the fore in interlinked ways. The first of these periods was the interwar years, especially from the mid 1920s until just before World War II. It was during this time that the British industry transformed itself from a craft-based industry into a factory-based one, adopting automated equipment for large-scale production that borrowed elements from Fordist mass production and from Taylorist "scientific management." New products, production methods, and management techniques were tried out during this period, while established methods were questioned and in some cases abandoned. The industry began to consolidate around a small number of large manufacturers through buyouts and mergers. At the same time, the bicycle market was growing rapidly. Bicycling both for utility and for leisure was at its peak. The second period I will focus on spans the 1980s and the 1990s. A steady decline in the cycling market during the preceding two decades had been matched by a near collapse of the British industry. Again, a number of factors came together at this point to revive both the industry and the market. The emergence of environmentalism and health consciousness as spurs to cycling coincided with an innovation—the mountain bike—that caught the imagination of the "baby boomer" market, a market with considerable disposable income. Production innovations were again crucial to the growth of this market, but more important was a fundamental transformation of how the industry was organized on a global scale. As a result, the British cycle industry is very different today from what it was in the past.

The focus on production makes this book a meeting point for a number of intellectual and political perspectives, spanning topics including the labor process, economic history, innovation studies, and organizational studies. Central to the discussion are debates on Fordism and what followed it (most commonly labeled post-Fordism) and sociological and cultural analyses of modernity, postmodernity, and globalization. One of the book's objectives is to bring the insights of these various approaches together within a critical analysis rooted in the sociology of technology. In drawing such links among technology, society, and culture I hope to throw some new light onto processes of technological change, especially the possibilities for directing change in ways that make the control and the accessibility of technology more egalitarian for producers and consumers, for designers and users, and for employers and employees.

I hope this study will also offer insights that might benefit the causes of radical social and political actors such as the environmentalists and transportation activists who champion bicycles. The roots of such perspectives lie within a further strand of thinking about technology—a strand that is linked to the counterculture and the alternative-technology movement of the 1960s and the 1970s and also to more philosophical interrogations of the politics of technology (e.g. Mumford 1975; Winner 1986; Sclove 1995; Feenberg 1999; Martin 1999; Kleinman 2000). These kinds of critical engagement with technology are highly pertinent in a context where bicycle advocacy has to engage with a politics and a culture of transportation that take the dominance of the automobile for granted.

Much of the book draws on archival material that is only just beginning to be systematically analyzed by researchers. The research was carried out between 1991 and 1994 and was revised after supplementary research in 1997 and 1998. The four main components of my research were (1) exploratory interviews with mountain bike owners, since the project began as an investigation of the technological and cultural changes that had occurred with the development of mountain bikes, (2) more strategic interviews with "key personnel" in the bicycle industry and the cycling culture, (3) observations of bicycle-related gatherings, notably several public and trade cycle shows, and (4) documentary research. The documentary research drew on several sources, most notably the Raleigh Cycles archives held by Nottinghamshire County Council, other archives pertaining to Raleigh held by Nottingham Local Studies Library, and cycling magazines (dating back to the 1880s but mostly from the 1980s and the 1990s). A complete list of the interviewees is given in appendix A, and the archival sources are listed in appendix B.

The perspective from which the material was collected and analyzed mixes together a variety of sociological and anthropological approaches, especially as they have informed the multi-disciplinary field of *science, technology, and society* (STS) and its subfield *social studies of technology* (SST). This reflects both my interdisciplinary background (veering across anthropology, cultural studies, sociology, and STS) and the multi-dimensional nature of my subject matter. Many academic disciplines can claim insights into the history and culture of cycling and cycle production. Research on different aspects of this topic has spanned technology design (Hult 1992; Roy 1983, 1984), economic history (Harrison 1977; Millward 1990, 1995; Lloyd-Jones and Lewis 2000), design history (Oddy 1994, 1995), social history (Ritchie 1975; McGurn 1987, 1999), and social geography (Patton 1995). The large subculture of mostly amateur cycle historians adds another dimension—especially since STS itself began to pay attention to the bicycle, with the publication of Pinch and Bijker's (1984) analysis of late-nineteenth-century cycle design. This has drawn a mixed response from academics (Russell 1986; Winner 1993; Rosen 1993) and cycle historians (Oddy 1995; Ritchie 1995; Clayton 1999). This book thus involves conflicting disciplinary perspectives, on which I draw as seems appropriate for each of the components of my argument. Underlying the eclecticism, though, is a commitment to understanding technological change in a way which is neither technologically nor socially determined, which pays attention to the contingencies and uncertainties of change, and which treats critically both the relationship between technology and society and the rhetorics of those who promote change.

Technology and Culture in the Project of Modernity: From the Manifesto of the Communist Party to the Manifesto for Cyborgs

A central theme underlying my argument will be the need to take seriously within the sociology of technology issues and debates from other intellectual fields. Most prominent among these in my discussion will be debates (which have become central to the sociology of culture) concerning modernity, postmodernity, and globalization. Accounts of these phenomena concern themselves with questions about the nature of social and/or cultural change, with how such change is constituted, and with what it means to the people who experience it. Technology, or rather technological change and innovation, is central to many of these accounts. However, what it is that constitutes technology for theorists of modernity and postmodernity is rarely articulated in much depth. Technology might be regarded instrumentally, as something used for specific objectives by the people or classes who are seen to have brought about major social and cultural change, or it might be treated as a social actor in its own right, with an intrinsic logic propelling society into an uncertain, dangerous, yet exciting future. More often it is regarded somewhere between these two poles. Technology is seen to bring about social change at the same time as it propels such change in specific directions. Rarely, though, do such theorists look more closely at technology, at the detail of how interactions among the social, the cultural, and the technical shape both artifacts and the processes of change. The detail of technology is, rather, usually incidental to the main arguments of such writings, even if its absence would make those arguments untenable. Despite the centrality of technology to the shaping of modernity, then, there has been a tendency to assume, in the words of Langdon Winner (1977: 2), that "the true problems of modernity could best be understood in ways that excluded all direct reference to the technical sphere."

Modernity and Postmodernity: Experience, Identity, and Culture

Marshall Berman (1982: 13) describes modernity primarily as an experience by which people "are moved at once by a will to change—to transform both themselves and their world—and by a terror of disorientation and disintegration, of life falling apart." He identifies this experience in writings by Goethe, Marx, Dostoevsky, Baudelaire, and others, linking it also to certain urban landscapes—in Paris, St. Petersburg, and most notably in the New York planning "vision" of Robert Moses. Berman goes on to describe "the highly developed, differentiated and dynamic new landscape in which modern experience takes place" (ibid.: 18–19):

> This is a landscape of steam engines, automatic factories, railroads, vast new industrial zones, of teeming cities that have grown overnight, often with dreadful human consequences; of daily newspapers, telegraphs, telephones and other mass media, communicating on an ever wider scale; of increasingly strong national states and multinational aggregations of capital; of mass social movements fighting these modernizations from above with their own modes of modernization from below; of an ever-expanding world market embracing all, capable of the most spectacular growth, capable of appalling waste and devastation, capable of everything except solidity and stability.

People experiencing this new world are, for Berman, caught up in the sense of living in a revolutionary age, in "the sense of being caught in a vortex where all facts and values are whirled, exploded, decomposed, recombined" (ibid.: 121).

A similar picture of modernity is painted by David Harvey, whose interest goes beyond Berman's primary concern with literature (and Berman treats even Marx's writings more as literature than as any other form of discourse). Harvey draws on geographical concerns with space and time, and on organizational and sociological questions concerning economics, production, and culture. Harvey goes beyond modernity to include its extension into "postmodernity," taking as his starting point Baudelaire's comment (quoted in Harvey 1989: 10) that "modernity is the transient, the fleeting, the contingent; it is the one half of art, the other being the eternal and the immutable." Like Berman, Harvey pays most attention to the first half of this formula, which he sees as even more important in the shaping of postmodernity than it was in the shaping of modernity. He examines the cultural episodes of new movements in the arts that have constituted "modernism" and "postmodernism," and he links them to the compression of space and time that have come about as a result of political and cultural change and to technical and social innovations such as those described in the quotation from Berman. He then traces these changes back causally to specific moments of overaccumulation and crisis in the economic sphere, which he argues were responsible for sparking both modern and postmodern forms of cultural representation.

For Harvey, then, the experience of modernity and postmodernity is simultaneously the experience of capitalism, whose inherent tendencies toward crisis lead not just to economic but also to social, political, and cultural upheaval. The depression of 1846–47 in Britain thus had multiple effects—it resulted in a crisis of representation that was manifested in the revolutions that swept across Europe in 1848 and in the publication of the Manifesto of the Communist Party (Marx and Engels 1977); it led to the emergence of new systems for organizing stock and capital markets; and it sparked new forms of art and literature that began to address questions of internationalism, synchrony, temporality, and economic exchange (Harvey 1989: 261–263). Similar crises in the 1960s and the 1970s have generated a further compounding of these features, resulting for Harvey in the emergence of postmodernity. He locates the distinction between modernity and postmodernity in a radical shift in politics, economics, and the experience of time and space. He sees as crucial to this shift certain changes in production and economics—from Taylorism and Fordism in the early to mid twentieth century to a globalized system of flexible accumulation along with related changes in the political sphere. While he makes it clear that he regards the shift to postmodernity as having been generated by the same crisis tendencies that precipitated modernity, it is the intensity of these tendencies that make the postmodern qualitatively different.

Zygmunt Bauman (1992) and Mike Featherstone (1991), in their discussions of the shift from a production-centered culture that goes with Fordism to a post-Fordist, postmodern consumer culture, pay more attention to the cultural rather than the economic practices associated with these changes. For Bauman (1992), this consumer culture is a social system that has replaced the modern producer culture, leading to a decentering of work in the industrialized West. This dynamic of production and consumption is central to modernity-postmodernity debates, since the shaping of Western culture in the twentieth century and the development of mass and then more flexible approaches to production have been closely intertwined. One important dimension of this, which is too often neglected in the enthusiasm to explore "consumer culture," is the geographical specificity of the shift from production to consumption, from industrial to "post-industrial" society. Notable here is the ubiquity of the concept of "globalization," which has become significant as a way of accounting for how the spatial aspect of postmodernity has taken on the same significance that modernity attached to the temporal (Featherstone et al. 1995). A strong focus on the globalized and somewhat homogenized consumption of cultural products means, however, that the more visible differences between the relationship of different regions and populations to production is often missed. As the production of goods in the West has declined in favor of growth in knowledge-based service industries, the job of meeting the demands of increased Western consumerism has fallen increasingly on producers in the developing countries and in the newly industrialized countries of the Pacific Rim (Barbrook 1990; Lipietz 1992). For people in these societies, it is modernization and not postmodernization that describes their current experience, yet this is a significantly different kind of modernization than that which has characterized the West since the sixteenth century (Featherstone et al. 1995; Seabrook 2000). It has emerged as a consequence of global transformations of industrial capitalism, and of the modern experience, but people in developing countries are at the receiving end of these processes and are not, for the most part, their initiators (Friedman 1995; King 1995).

There is a danger, then, of overlooking the complexity of the relationship between production and consumption by focusing on only one or the other. This has been a common theme of critiques of more traditional production-based analyses of modern society. Baudrillard (1988) criticizes Marx and the Marxists who focus solely on production, neglecting the socially mediated construction of use and of need and neglecting the role of consumption as an integral element of the production process. Similarly, Johnson (1986–87: 55) criticizes the "productivism" of Gramsci, in which the cultural life of a product is seen to be determined by the conditions of production rather than its interplay with consumption. This interplay is increasingly being explored in work produced at the borders of cultural studies and technology studies. In contrast to the sharp distinction drawn by earlier writers between production and consumption, many are now increasingly concerned with the active work of consuming. Consumers don't just blindly accept the output of producers; they make creative choices about what they consume—choices linked especially to the establishment of consumer identities (Miller 1987; Du Gay et al. 1996; Silverstone and Hirsch 1992).

The construction of consumer identities is a concern of producers, too. Concepts such as "champions of innovation" and "early adopters" are used in the literature of business studies to refer to customers with "needs" ahead of the rest of the market, who can provide information on potential new avenues for innovation (Bailetti and Guild 1991). The feedback of early users of new products can also be valuable as a guide in refining the design of subsequent models. This is frequently the case with information technology (Skinner 1992), and it has been documented with regard to consumer products too (Cockburn and Ormrod 1993; Akrich 1995; Du Gay et al. 1996). While these kinds of processes can—from the perspective of producers—facilitate attempts to shape consumer identities in ways that can have commercial benefits, they are also, of course, open to resistance and contestation (Woolgar 1991; Akrich 1992). What they show is that production and consumption are multiply interwoven in ways that were central to the dynamics of postmodernity in the late twentieth century.

Technological Change and Cultural Change

If modernity was sparked by a combination of the Enlightenment, industrial capitalism, and European colonialism, it could not have happened without the technological changes that accompanied these episodes. However, despite their sophisticated analyses of cultural change, accounts of modernity and postmodernity rarely subject the relationship with technology to any serious scrutiny. Harvey's 1989 account of postmodernity explores the mutual shaping of technological and social change in relation to Fordism and the emergence of flexible specialization. However, his broad-brush approach, relying on national and international statistics combined with fairly sweeping statements about the nature of flexible specialization, prevents any consideration of technical detail. More problematic is how Lyotard (1984) and Baudrillard (1988) treat as given technologies that they cite as central to the postmodern—notably, computerization and television—without (to paraphrase Mulkay 1979: 80)[2] showing what it is that makes these technologies postmodern in the 1980s when earlier they would presumably have been just modern. A teasing out of how the emergence of (post)modern society has been shaped by and has subsequently shaped technological developments is largely absent.

One exception is the work of Donna Haraway, whose "manifesto for cyborgs" (1989) takes a broad and unremittingly political approach to the hybrid, multi-dimensional nature of technology. Haraway (ibid.: 178) writes about "the informatics of domination," the networks of new information technologies that have serious yet hidden implications:

> Our best machines are made of sunshine; they are all light and clean because they are nothing but signals, electromagnetic waves, a section of a spectrum, and these machines are eminently portable, mobile—a matter of immense human pain in Detroit and Singapore. People are nowhere near so fluid, being both material and opaque. Cyborgs are ether, quintessence.
>
> The ubiquity and invisibility of cyborgs is precisely why these sunshine-belt machines are so deadly. They are as hard to see politically as materially.

Haraway directly links globalization to technology in postmodernity. She points out that production is as likely to take place in the Pacific Rim as in the West, with the same poor conditions for workers. Marx responded to bourgeois capitalism's globalizing tendencies (see Marx and Engels 1977) with an equally "grand design," the call for "workers of all countries [to] unite!" Instead of this, Haraway adopts what she calls the "cyborg myth" as a postmodern form of resistance that matches the less tangible movements of global capital in postmodernity. However, like Marx, she portrays a two-sided vision, populated by both oppressive and resistant cyborgs. Her vision shares with other feminist utopias (e.g., Piercy 1979, 1992) an ability to see technology as simultaneously liberating and threatening, though in essence neither. The sociotechnical trajectory of an oppressive social structure might yet develop in a different direction—it is always open to the disempowered to try to bring about change, and no outcome can be taken for granted as certain.

The Politics of Technology Studies

Haraway's desire to promote an alternative vision to a technologically based authoritarianism is matched elsewhere in a wide range of technological critiques and visions. Within STS itself, an early component of the field—alongside the parallel "radical science" movement (Levidow 1986)—was a critical account of the way science and technology were developing during the Cold War, linked to the peace and environmental movements of the 1960s and the 1970s (Cutcliffe 1989; Bijker 1993). The STS principle of deconstructing established notions of what constitutes science and technology is highly compatible—in principle if not often in practice—with a number of critical strands of technology analysis. In particular, Mumford (1975), Ellul (1964), Illich (1973), Winner (1977, 1986), Sclove (1995), Bookchin (1974), and Feenberg (1991, 1999) have focused on questions about how well different kinds of technology facilitate democracy, autonomy, and community. Fundamental to these analyses are questions as to who authorizes the development of an artifact or a technological system, who then decides on its design, and how this then informs the relationship among policy makers, innovators, and "the public" (Sclove 1995; Joss 1999).

Such concerns also bear on practices within STS, where political commitment and outcomes have been topics of ongoing debates. Pinch and Bijker's constructivist approach to technology—the Social Construction of Technology (SCOT) approach, which I will be adapting later in this chapter for use as the basis for my own analysis—has been criticized by Stewart Russell (1986: 335–336) for not addressing the political relations among the "relevant social groups" that are seen to shape technological meaning, and for not exploring these groups' "differing abilities to influence the outcome" of technological development and adoption. These concerns are reiterated by Winner (1993: 440–442), who argues that constructivist accounts of technology unwittingly serve the interests of the powerful by asking only how the meanings associated with an artifact become stabilized, and not why this occurs or to whose benefit. By extension, Winner regards constructivism as narrowly academic in its concerns because he sees it as refusing to take a political standpoint on technological issues.

In fact, a concern with the political dimensions of technology has been central to STS since the field was established. This concern has, though, only occasionally filtered through into work that has, like SCOT, followed the tradition of the sociology of scientific knowledge (SSK), as opposed to other disciplinary components of STS. The work of Brian Wynne (1988) is a notable exception, bringing together approaches from SSK with critiques of technology assessment, risk assessment, and other aspects of public policy, particularly where this involves a clash between institutional and "lay" knowledges and interests. The politics of SSK has itself become subject to debate more recently (Ashmore and Richards

1996), while other dimensions of STS—studies of technology and work, and technology and gender, for example—have served to redress some of the political shortcomings of the field (MacKenzie and Wajcman 1999; Noble 1984; Wajcman 1991).

There is, nevertheless, some justification for Winner's argument that broader political questions can get lost in programmatic accounts of how specific artifacts come to be socially constructed. Pinch and Bijker's (1984) account of the relevant social groups in the world of nineteenth-century cycling pays little attention to the class aspects of who was able to afford a bicycle at that time, or to how this changed between the 1870s and the 1890s; nor does it address, despite discussing the role of women in shaping the meaning of bicycles, the position of women in British society, or how this compared with that of French women (McGurn 1999).[3] From a slightly different angle, both Bijker's (1992) analysis of fluorescent lighting in the 1940s and Callon's (1986a, 1987) analysis of electric vehicles in the 1970s miss the opportunity of exploring the role of "new social movements," notably environmental ones, in raising new political questions about technology. In the 1980s and the 1990s, the "high-intensity lamp"—whose stabilization in the 1940s is described by Bijker—was reconstructed as a "low-energy" lamp that met the ecological requirements of reducing carbon dioxide emissions from electric power stations. Similarly, the search for ways to develop satisfactory electric vehicles has received a new stimulus from the changing international policy agenda concerning pollution and sustainability, also raising questions about individualistic transportation policies.

Developing from this kind of problem with STS studies, a focus on the minute details of design alongside only the narrowest of social contexts means that many studies fall short of asking the kinds of questions that Winner and many feminist analysts would want to ask about technological change in relation to the kind of society we wish to live in. In other words, these studies rarely question the basis on which specific technologies are developed in the first place. Military technology is an important case here. Law and Callon's (1992) account of the TSR.2 aircraft, for example, traces how the relationships among the local and global networks of government, defense agencies, and the aerospace industry shaped the construction of this sociotechnology. As Mort and Michael (1998) note, a different account might have questioned the commitment of various actors in the story to a defense system based on nuclear weapons, or explored the implications of the tendering process for the workforces of the two companies engaged to build them. (See also MacKenzie 1990.)

Despite these misgivings about certain aspects of technology studies, I do not concur with Winner's argument (1993: 449) that themes running through the sociology of technology—such as the insight "that the course of technological development is not foreordained by outside forces, but instead a product of complex social interactions"—are "increasingly redundant." While studies such as those just discussed often fail to make broader political and ethical connections, much of what constitutes "the sociology of technology" is compatible with a more than purely descriptive analysis (Feenberg 1999). Bijker (1993) and Grint and Woolgar (1997) argue for a more politically relevant sociology of technology, although they do not quite manage to achieve this objective in these particular books. Nevertheless, this is not to negate the "deconstructive capacity" of constructivist sociology of technology "to show interpretative flexibility, to suggest alternative technological choices, to debunk the sociotechnical ensembles constructed by the powerful" (Bijker 1993: 130).

The Social Construction of Technology

How can constructivist technology studies fulfill its "deconstructive capacity"? A widely stated objective within SST is to take seriously the detailed technical content of particular artifacts, but at the same time to locate this detail within a social and cultural context (Staudenmaier 1985). This perspective profoundly challenges the notion of any inherent technological trajectory that might direct technological change toward a supposedly inevitable end point (MacKenzie and Wajcman 1985; Winner 1977). Rather, technology is presented as something contingent and emerging—something that in different circumstances "might have been otherwise" (Bijker and Law 1992: 3).

At the same time, as events unfold in the story of any particular technological artifact or system, SST analysts recognize that the technological and social paths that are established do begin to solidify such that they become less easy to dislodge. In other words, anything could happen in the earliest days of an innovation, but this becomes less and less so as time goes on. As the cultural meanings associated with an artifact stabilize alongside its technical features, it becomes less malleable and more fixed. The configuration of technology and society within an artifact thus becomes more resistant to change—more obdurate (Bijker 1995). Social studies of technology resist, then, treating technology as given—it is regarded as part of a seamless web that also includes society, culture, politics, economics, "etcetera, etcetera" (Hughes 1986). How these different elements come together to form sociotechnical ensembles (Bijker and Law 1992) which are embodied in specific artifacts deserves further study, and this also holds the key to unpacking the politics embedded in technology in ways that can benefit those at the margins. Yet this is made more difficult by the fact that, in a reversal of the problem with cultural theory, while aiming always to problematize technology, SST accounts rarely give "society" the same treatment. Social studies of technology tend to restrict their conception of the social to the immediate social context of a particular artifact. A typical contextual study will focus on the social circumstances surrounding a technology's invention and diffusion—for example, significant events involving the economic, political, and/or social relations among companies that result in the emergence of one specific configuration of that technology rather than another. It is less common in artifactual case studies to locate technology within a broader context of social or cultural change, particularly in relation to change at a "macro" level (Feenberg 1999). Reversing this trend requires, I believe, some adjustments to SST theory. My conceptual approach in this book draws on one particular version of SST: the SCOT framework, developed by Trevor Pinch and Wiebe Bijker (1984; see also Bijker 1995). I will first sketch the basic SCOT framework, then highlight a number of shortcomings (shared in many cases with other SST approaches too)—in particular, an imbalance between the technical and the social or cultural, and a focus on discrete artifacts. This can be problematic in the case of a messier setting in which several different artifacts are equally significant. Despite these problems with SCOT, there are also benefits to using this framework; I will therefore go on to sketch out an alternative way of approaching the social construction of technology. Based on the concept of sociotechnical frames, this conceptualization allows, I believe, a better articulation of the political dimensions of sociotechnical change—and hence a better chance of contributing to a more radical political agenda for technology.

SCOT is actually a developing framework, having been refined and modified since the mid 1980s—especially by Bijker and Pinch themselves, but also by others (Bijker and Bijsterveld 2000; Aibar and Bijker 1997; Kline and Pinch 1996; Elzen 1986; MacKenzie 1990; Misa 1992; Blume 1997). The essence of the approach is that technology is understood to be constructed not just by means of engineering but also by the other activities of engineers and of others associated with an artifact. The construction of technology is thus simultaneously social and technological, meaning that technology should be more accurately labeled sociotechnology (Bijker 1995). Sociologists of technology using the SCOT approach have endeavored to trace how it is that certain configurations and not others of social and technical elements combine to form particular sociotechnical ensembles (Bijker and Law 1992). For Bijker, for Pinch, and for others, understanding this process is achieved by means of a theoretical toolbox that includes four main concepts: relevant social groups, interpretative flexibility, closure, and stabilization. These concepts cohere around a theoretical structure that Bijker terms a technological frame, in relation to which actors may experience a greater or a lesser degree of inclusion.

The heuristic of the relevant social group plays a central role in SCOT as a guide in tracing sociotechnical change. Relevant social groups are a means of "following the actor" (Latour 1987), of understanding technology from the inside; for Bijker, focusing on the problems and solutions each relevant social group attaches to an artifact is the key to understanding the fluidity of technological change. Competing interpretations of an artifact by different relevant social groups result in interpretative flexibility. In other words, something that is a successful technology for one group may be a serious failure for another, and both meanings can coexist for the same artifact. Consequently, there is no clear way of defining technological "success" or "failure," and Bijker stresses his refusal to explain the development of an artifact in terms of a retrospective reconstruction of success—a common feature of non-constructivist, linear, accounts of technology. Such an account would claim that today's bicycle, refrigerator, or computer is the pinnacle of technical development, the "natural" successor to earlier versions. SCOT shows that such is rarely if ever the case.

SCOT explains the elimination of interpretative flexibility by means of closure and stabilization. Bijker's own distinctions between these two concepts is at times slippery, but it seems most useful to see closure as what occurs when a consensus emerges among all relevant social groups in regard to an artifact's dominant meaning, allowing alternative meanings to fall into disuse (Bijker 1995: 86). This can be brought about by a number of different closure mechanisms (Pinch and Bijker 1984; Beder 1991). Bijker (1995: 271) argues that the process of closure is "(almost) irreversible." Its irreversibility explains for Bijker what he terms the "obduracy" of sociotechnical ensembles—that even though technologies might have been otherwise at one point, once the interpretative flexibility of an artifact has closed it is extremely difficult to revive former meanings. Once closure has occurred, "stabilization" can proceed. By this

Bijker generally means the gradual process by which the technical characteristics of an artifact become standardized and begin to be taken for granted—in other words, closure is achieved materially as well as interpretively. He draws on Latour and Woolgar's (1979) use of modalities attached to statements about scientific facts as a means of ascertaining how stable an artifact has become—in other words, an artifact is stable when no qualifying terms are needed in order for somebody to understand what is being referred to (Bijker 1995: 86–87).

This set of theoretical building blocks, used by Bijker and others to explain technological change, is underpinned by the structural concept of technological frames, which, for Bijker, are constituted through interactions concerning particular artifacts, and which develop alongside relevant social groups. If there are no interactions concerning an artifact, there will be no technological frame and no relevant social group (Bijker 1995: 123). This allows Bijker to move away from the myth of the individual inventor genius producing an artifact by himself (or, rarely, herself) in the laboratory or the workshop, instead locating the act of invention within a social context that allows both continuity and change in the development of technology. This is achieved in particular via the notion of inclusion. An actor can be simultaneously a member of several relevant social groups, and involved in a variety of technological frames, but the degree of his or her inclusion in each frame will vary. Greater inclusion in a frame will mean that an actor's activities are highly structured by it, and they will frequently draw on it as a resource (ibid.: 143). Where they have a lower degree of inclusion, this will be less the case. Bijker suggests that significant episodes of technological change often come about when a significant actor shifts his or her primary association from one frame to another and, at the same time, brings to the new frame methods and approaches from the first frame. In this way, technological frames provide a way in which stability is maintained among a community of actors sharing an interest in particular artifacts; at the same time, they allow a structured understanding of sociotechnical change. This theoretical approach offers many benefits to those trying to understand the interaction of technological, social, and cultural change—to understand sociotechnical change. It offers the ability to span the risky dichotomy between agency and structure by providing a framework that recognizes the influence of a cultural milieu on the inventors who push technological change forward. It takes account of the myriad social groups that influence how change develops, and the complex ways in which these groups interact. And it tries to understand how a drive toward change can exist alongside a much greater degree of technological continuity.

However, SCOT is still an emergent theory, not yet fully and consistently developed, as Bijker himself recognizes. Most prominent among its shortcomings is that it focuses primarily on technical concerns. Crucial here is what Grint and Woolgar (1997) characterize as *technological essentialism*—the common assumption that there is a clearly identifiable essence that underpins technology and its effects on society. If technologies develop in ways that are contingent and context-dependent, there can be no "true" technological essence. Grint and Woolgar argue, therefore, for a post-essentialist position that moves beyond mere anti-essentialism by resisting the technicist assumptions frequently included—often unwittingly—in many anti-essentialist writings. As they recognize, it is difficult to develop a consistent anti-essentialist standpoint while still holding onto some level of technicism, a problem that is visible in Bijker's (1995) main exposition of SCOT. As Grint and Woolgar might predict, despite Bijker's strong and convincing argument that technology should be treated as more than just technical, he and other SCOT proponents still fall short of fully integrating within their analyses the non-technical aspects of technology along with the technical. This criticism applies both to the many case studies of technological change that have been published and to the way SCOT separates stages of analysis first along technical lines and then along social lines (Pinch and Bijker 1984; for a critique of this, see Rosen 1993). For example, Bijker initially (1987: 168) described a technological frame as "a combination of current theories, tacit knowledge, engineering practice (such as design methods and criteria), specialized testing procedures, goals, and handling and using practice". Later (1995: 125), Bijker refined and extended this definition into the following list of elements:

goals
key problems
problem solving strategies
requirements to be met by problem solutions
current theories
tacit knowledge
testing procedures
design methods and criteria
users' practice
perceived substitution function
exemplary artifacts.

This list includes features that Bijker regards as applicable to all relevant social groups—not just those of engineers—on the ground that "all social groups should a priori be treated as equally relevant" (ibid: 123). Thus, while its components derive primarily from the concerns and from the language of engineers, Bijker wants them to be interpreted in relation to other relevant social groups, and some indeed are specifically related to users rather than to designers and builders. Nevertheless, despite Bijker's clear commitment to a notion of sociotechnology, it isn't clear what the place of society is within a technological frame, except for the very specific society of specialist engineers working in the relevant field.

Another gap within SCOT is the cultural dimension of technological change. Beyond the narrowly defined problems and solutions pertaining to a particular artifact, there are wider cultural factors at play which also contribute to its shaping but which could be lost sight of in this framework. Mulkay (1979: 95) writes how scientific knowledge is established "by the interpretation of cultural resources in the course of social interaction." For Mulkay, scientists draw on both the cultural resources of their narrower scientific culture and the cultural resources of the wider society in which they live. This explains "the dynamic social processes whereby science absorbs, reinterprets and refurbishes the cultural resources of modern industrial societies" (ibid.: 121). This notion can be applied equally to technological and to scientific practices, and it is valuable in making clear the equal importance of diverse cultural influences on the development of technology. The more specific "cultures" that take shape around specific artifacts also should be borne in mind. These "cultures" include activities, events, and publications that promote and support the consumption of an artifact; they also include the cultures, practices, and narratives of the various organizations producing, promoting, and using the artifacts (McLaughlin et al. 1999).

It is important to recognize the distinct role played in sociotechnical change by the users of an artifact. Users' concerns appear to be treated within Bijker's technological frames as secondary to those of designers and engineers, and usually just in relation to the early stages of the product life cycle—primarily conception and design. As I have already discussed, though, users influence both the social and the material construction of an artifact throughout its life cycle, both by feeding back to manufacturers ideas that can be incorporated into the next model and by appropriating artifacts within their own systems of meaning in the process of consuming them (Skinner 1992; Silverstone, Hirsch, and Morley 1992; McLaughlin et al. 1999). More fundamentally, users' practices should be understood as concerned not just with artifacts themselves but also with broader strategic and discursive objectives. These can change as the relationship between user and artifact changes, and they can vary between different groups of users who might adopt different strategies in relation to it. Thus, discursive practices might further our understanding of sociotechnical change.

Another significant problem with Bijker's approach arises out of the fact that he applies the concept of technological frames only in relation to discrete artifacts, specifically Bakelite and lamps. Had he instead set his study of high-intensity fluorescent lamps within the context of the broader electrical industry, or had he revisited the bicycle industry, he would have found it difficult to apply this framework without some revision. In particular, it is not easy to go on applying the notion of technological frames only to discrete artifacts when faced with a more diverse group of machines and tools that might bring up somewhat different design problems and objectives but which nevertheless share many other features and can only be understood meaningfully in relation to each other. Examples would include a range of separate but related products of the same industry, the changing features of what is nominally the same artifact over a period of time, or a group of different artifacts that play complementary roles within a single technological process. These possibilities raise the question of what actually counts as "an artifact" for SCOT, in terms of there being a discrete set of features that constitute a distinct technological frame around which different relevant social groups can mobilize.

An analysis of sociotechnical change that is focused on production will have to account for a range of practices and activities that cut across social groups, across artifacts, and across moments in the life cycle of any specific artifact—an empirical setting that lies beyond the analytical capacity of Bijker's framework. The difficulty faced by Bijker's version of SCOT in addressing such a situation resembles a problem Grint and Woolgar (1997: 126) identify in the sociology of the workplace: ". . . just as most sociology of work has turned out to mean the sociology of the factory assembly line so, most of it has turned on the production of artefacts through technology rather than the consumption of technological artefacts." Here Grint and Woolgar refer especially to the organizational consumption of production technology as opposed to individual consumption of the artifacts being produced. (See also McLaughlin et al. 1999.) A parallel though partially inverted problem presents itself in the sociology of technology, which concerns itself almost exclusively with the design and consumption of artifacts and pays little attention to their production or to the technologies of production. With Bijker's conceptual framework, it is easy to see why this problem exists. The framework he uses works well when applied to products that can be conceived as fairly discrete artifacts caught for the most part in exchanges between producers and consumers—albeit artifacts in a state of interpretative flexibility. But he does not bring into the frame other artifacts that interact with these products within the sphere of production.

The move from the design workshop and the marketing department to the factory floor presents problems for SCOT. Here the products being manufactured encounter other technological artifacts, notably the production equipment used to make them. These two kinds of artifact are each essential for understanding the other, since they have been mutually shaped—products by the nature of the production equipment used, equipment by the specific requirements of product design. There is, though, no way of accounting for their interdependence within the context of a technological frame, which has space only for separate and discrete artifacts. Similarly, the shop floor complicates relations among social groups—we are no longer dealing only with competing designers, communities of engineers, and users and non-users of an artifact. Rather, the production of goods involves a wide variety of staff categories within an organization, differentiated both within and between the workforce and management. This situation renders even less sustainable the engineering bias of the elements that Bijker sees as constituting a technological frame.

In fact, Bijker makes no definitive claims for his theory—indeed, he considers his list of features of a technological frame to be tentative, empirically generated, and open to change (1995: 125). It is this openness that makes it appropriate to try to adapt and improve SCOT in order to overcome its shortcomings. What I want to do here, then, is suggest how the components of the SCOT framework can act as stepping stones toward an approach that more adequately supports its own program of analyzing technology as socially constructed. This revised version of SCOT will then underpin my account of sociotechnical change in the bicycle industry.

Sociotechnical Frames and Sociotechnical Change

How can Bijker's framework be adapted to overcome the difficulties I have outlined? How can cultural resources, organizational practices, interlinked technologies, and the interactions between designers and users be incorporated into a technological frame? Even if this could be made to work satisfactorily, what status would they hold in relation to the elements already listed?

What I propose as an alternative framework for understanding the complexity of sociotechnologies is set out in figure 1.1. A sociotechnical

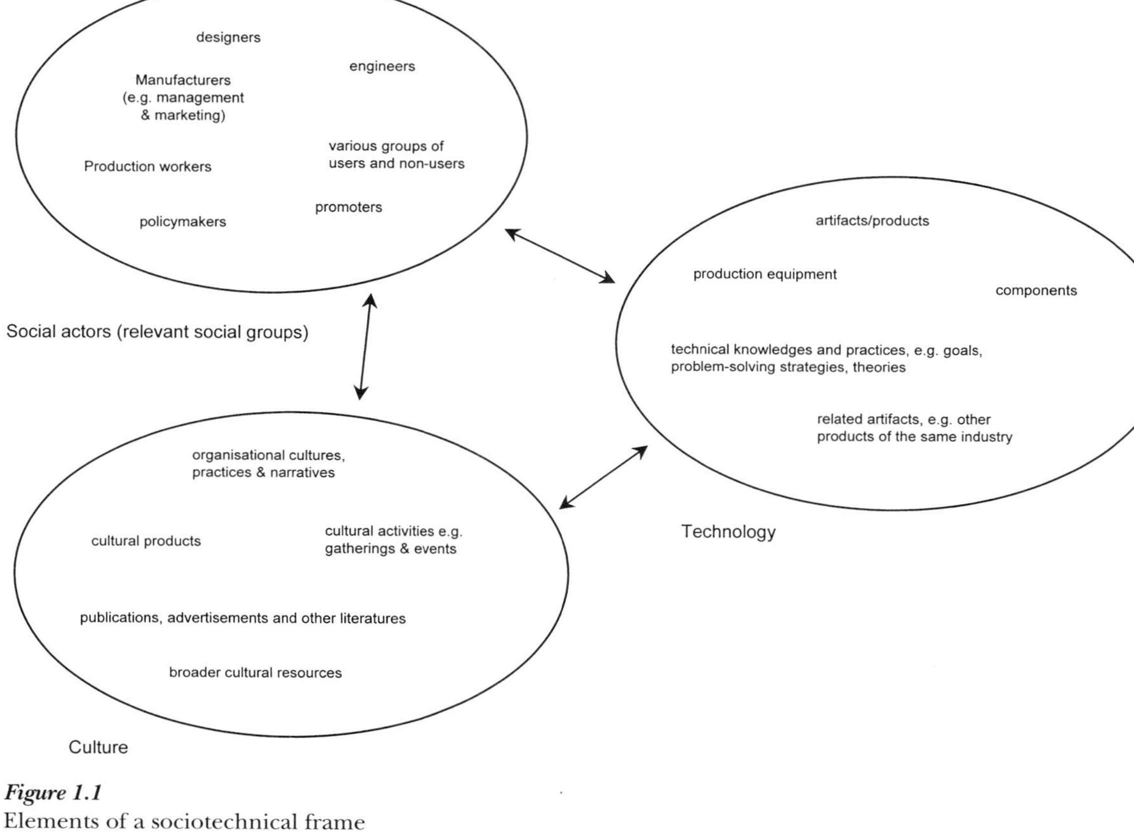

Figure 1.1
Elements of a sociotechnical frame

frame situates the technology of an artifact as just one of three components of the frame, taking us away from the narrower technical focus of a technological frame. The three components of a sociotechnical frame are (1) the social world of social actors—individual and organizational—associated with particular artifacts (Bijker's and Pinch's relevant social groups), such as designers, engineers, manufacturers (including groups differentiated along lines such as "management" and "labor"), promoters, policy makers, and various groups of users and non-users, (2) the technological world of artifacts and their components, process and production equipment, and other, related, artifacts, and (3) the cultural world that develops around artifacts in the form of organizational cultures, narratives and practices, gatherings and events, literatures, and other products, along with broader cultural resources on which various social actors can draw.

A sociotechnical frame is, in fact, something quite different qualitatively from Bijker's framework. His theoretical structure is primarily about the relations of engineers to their machines. The elements Bijker lists relate quite closely to the particular world views of communities of engineers associated with a specific artifact, even though it is intended not to exclude relevant non-engineers. Sociotechnical frames, in contrast, are not primarily about engineers and their machines. Rather, they structure the relations of a more heterogeneous group of actors and artifacts—not only relations between engineers and machines, but also relations between non-engineers and machines, between different kinds of related machines, and between different groups of actors (engineers and others). Mediating these relationships are the cultural values and practices that express the concerns of social actors and groups focused on a specific artifact or process. Sociotechnical frames thus encompass not only the elements of a technological frame (as outlined by Bijker) but also the groups of artifacts that have meaning for those involved, the significant events in the construction of the central artifact, and related technical processes and technologies.

How does this framework account for sociotechnical change? Bijker's framework offers two different models of change, set out in his case studies of Bakelite and fluorescent lamps. With the latter, he examines how a new artifact was constructed out of the conflicting interests of actors working within different technological frames—between manufacturers (whose goals and problem-solving strategies were geared toward selling lamps) and utility companies (whose major focus was to sell electricity and provide a public service) (Bijker 1995: 236–238). The result was a compromise between these two sets of goals that generated a new artifact and a new technological frame. (See figure 1.2.) In the construction of Bakelite, change in one technological frame was brought about through the work of an actor whose training and commitment lay within a different frame. A low level of inclusion within a technological frame can result, then, in an actor's finding radical solutions to a problem that are incompatible with established methods. This can then lead to the emergence of a new technological frame (figure 1.3). This conception of change as the outcome of tensions between technological frames is convincing, but it cannot provide an adequate explanation of how change is achieved across the messier interactions of the variety of technologies, actors, and events that constitute a sociotechnical frame. Changes to the meanings, the constructions, or even the material basis of one particular artifact will not necessarily bring about a transformation of the entire sociotechnical frame within which it is located. For that, it is necessary to look for instances in which the two processes Bijker describes are operating in tandem—that is, processes in which a clash between two frames takes place at the same time as marginal actors step outside the conventional frame, learn from

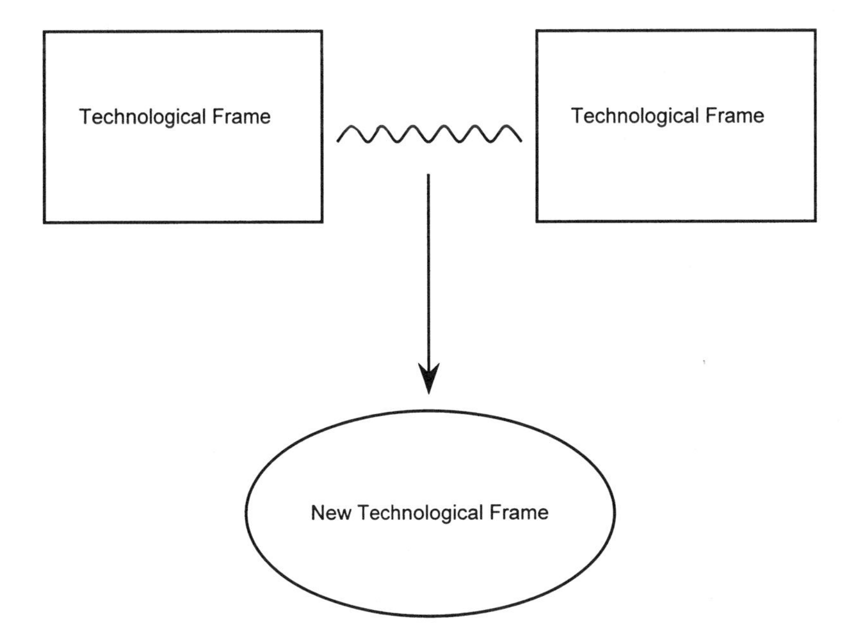

Figure 1.2
Bijker's account of how conflict between two existing technological frames leads to the establishment of a new one.

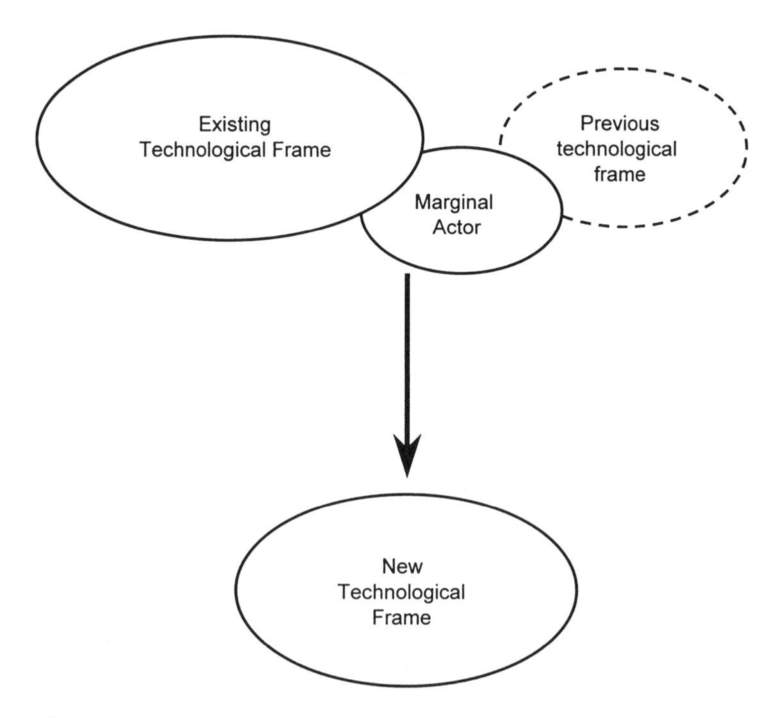

Figure 1.3
Bijker's account of how an actor with low inclusion in an existing technological frame establishes a new one.

alternatives, and thus develop a new frame (figure 1.4). The transitions that consequently take place from one frame to another hold the key to sociotechnical change—indeed, transitions between frames, rather than stability within them, are the focus of my empirical study. Such transitions come about when the three components of a frame (the social, the cultural, and the technological) get out of step with one another—more specifically, when the cultural component's mediating role between technology and society is no longer effective. According to Sharon Traweek (1992: 437–438), a scientific community is "a group of people with a shared past, with ways of recognizing and displaying their differences from other groups, and expectations for a shared future." "Their culture," Traweek continues, "is the *ways,* the strategies they recognize and use and invent for making sense, from common sense to disputes, from teaching to learning; it is also their ways of making things and making use of them and the ways they make over their world. . . ." The demise of a sociotechnical frame results when the social actors involved with a particular arti-

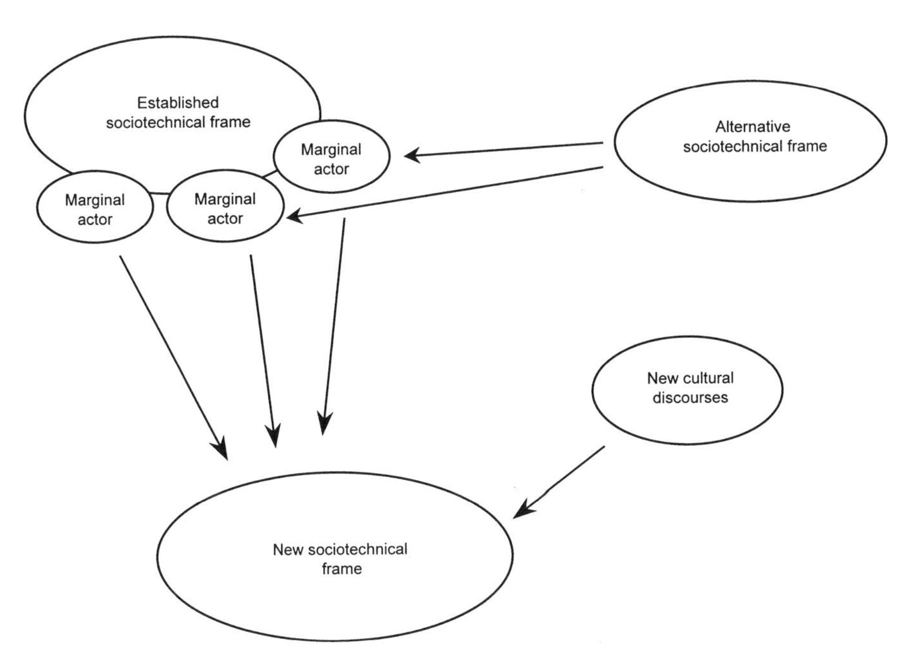

Figure 1.4
How marginal actors establish a new sociotechnical frame through encounters with alternatives.

fact no longer have these shared ways in relation to it. This might arise because cultural change has meant that different groups of actors no longer relate to each other or to the technology in the same way, or because the makeup of the social groups concerned with that artifact has changed, or even because the technology itself has changed (either through innovation or obsolescence) in such a way that it has lost its relevance for those social groups.

The value of using the concept of sociotechnical frames to understand this material is that it goes beyond the boundaries of the artifact and its immediate relevant social groups to account for a more complex interrelationship of technology, society, and culture. This then allows an understanding not only of change but also of stability. At a time when technology is being increasingly opened up to question, an important role of social studies of technology must be to examine closely the process by which technologies shift from being in a state of openness, of interpretative flexibility, to being fixed, embedded, stabilized, or locked into social structures (Nelis 1999; Bijker 1995; Winner 1977). How final is such fixing? Is the direction taken in any specific episode of technological change truly inevitable? And what scope does the process of sociotechnical change leave for movement away from technologies that are regarded as harmful—whether to the environment or to society—if, as Bijker writes (1995: 271), the process of technological closure is "(almost) irreversible"? Trying to find answers to such questions is crucial to achieving any kind of resolution in the politics of technology or in the politics of transportation and the bicycle.

The Politics of Production and the Politics of the Bicycle

How does this revised SCOT framework account for the unfolding story of the bicycle? In the following chapters I will apply this approach in describing the two significant changes in the British cycle industry: the transition from a sociotechnical frame of the factory bicycle to one of the mass bicycle and the transition from the latter to a frame of the globally flexible bicycle. I will then explore how constructivist technology studies might help precipitate another shift in the sociotechnical frame of the bicycle by promoting a participatory politics of bicycle technology.

This sociotechnical progression—from the local factory to a global arena—will be set against the background of wider changes in society, culture, and the politics of production. How did new forms of organizing the bicycle industry, its methods, its products, and its markets come about? What have the outcomes been? Where have such changes left British bicycle production and consumption? More generally, what does this story tell us about technological and organizational change, innovation, and the relations between technology and culture, between producers and consumers, and between manufacturers and their workers? Who are the agents of change, and how do their personal goals tie in with wider cultural values? How easy or difficult would it be to direct change in more purposeful, and hence less contingent, ways in order to achieve particular objectives? What implications does an analysis of industrial production have for social studies of technology? Such questions have barely been touched on in the cycling literature—or indeed in much other literature on technological or cultural change. Yet developments in bicycle design, production, and use were highly influential in the development of modernity from the late nineteenth century on. In the late nineteenth century, bicycles stimulated desires for personal transportation and for speed. They provided a private alternative to the public transportation offered by trains and horse-drawn coaches without the overhead or the infrastructure that these required. They shrank distances between villages and towns, breaking down boundaries that had previously defined the limits of communities, families, and working life and stimulating inventors to further refine the bicycle and to find ways of replacing human pedal power with other kinds of power. In addition to this role as a conceptual stepping stone toward the automobile, the nascent American bicycle industry of the late nineteenth century also served as a stage in the progression toward the mass-production auto industry that emerged just a few decades later.

Having played such a pivotal role in the great social, cultural, and technological upheavals of the late nineteenth century and the early twentieth, the bicycle has again become central to change. Long after its replacement by the automobile as the primary means of transportation in the developed countries, the bicycle remains predominant in many developing countries. Now, as those countries begin to look toward greater industrialization and an automobile-based economy, the bicycle is again being highlighted as the solution to the transportation problems of Western cities. Campaigners for "green transportation" and appropriate technology have long recognized the role that bicycles could play in making transportation more responsive to human needs and less harmful to the environment. As congestion and pollution levels increase, the car culture that has dominated Western social and economic life during recent decades is no longer regarded as sustainable. The alternative being promoted is an integrated approach to transportation, focusing on buses, trains, walking, and cycling.[4]

The changing fortunes of the bicycle and of cycling—varying historically in the West and also varying at any particular time in different geographical locations—exemplify the importance of understanding sociotechnical change within its social and historical context. They also point to the strong embedding of politics within technology. Political disputes inform the story of the bicycle at every level of analysis and at every stage in the life cycle of this artifact. In design, important questions arise as to who machines are designed by and for, and this extends also to the related practices of marketing, selling, and using bicycles. In the production process, the relations between the sellers and the buyers of labor power are as tense in the bicycle industry—and are just as subject to continual change—as in industries more commonly discussed in regard to the politics of production. Aside from these tensions between capital and labor, the bicycle industry as a whole has throughout its history been dogged by struggles among corporate rivals, the outcomes of which have affected the shape of cycling as an activity as well as the practices and processes of cycle manufacture. Which kinds of machines, components, and accessories become available to consumers is influenced by the technological choices set by these interactions.

Another aspect of the politics of the bicycle (and of transportation more widely) is the growing support for the environmentally benign transportation policies mentioned above. The increasing sense of urgency in solving problems of traffic congestion, pollution, and urban design—problems which are simultaneously environmental, social, and political—raises crucial questions about the control of space and about access to adequate transportation and to a healthy environment. How can we reverse the trajectory of our transportation system, which for 100 years has oriented the design of urban space, the organization of work and leisure, and the structure of the global economy toward the automobile? As I will argue in chapter 7, the bicycle has the potential to play a crucial role in such a reversal. For this reversal to happen, though, a great deal must change about the ways in which bicycles are produced and used, and especially in the meanings they hold for us. Most importantly, the prominent meaning of bicycles with which I opened this chapter, as a relic from childhood that adults might use only for leisure, will have to be superseded by a completely different conception of bicycles—a conception in which their use for leisure is secondary to their use as a mode of transportation.

2

The Emergence of the British Bicycle Industry

The history of the British bicycle industry has, for the most part, been simultaneously the history of the Raleigh Cycle Company. Raleigh was founded just at the period when small-wheel "safety" bicycles were beginning to take over from the high-wheel "ordinary" in production, in sales, and—most importantly—in people's conception of which design was best for speed, comfort, and safety (Dodge 1996; Bijker 1995). The company survived the industry's slumps of the late 1890s and the early twentieth century to emerge during the 1920s as one of the major players in British bicycle production. From this point until the 1950s Raleigh established a policy of taking high risks through the combined strategies of expanding and modernizing its factory and equipment—often during times of general economic hardship—while buying up many of its competitors as they each hit hard times. By the late 1950s, the cycle industry was entering into a decline, and Raleigh merged with its last major competitor, the British Cycle Corporation, which had been pursuing a similar expansion. TI Raleigh, the new company formed through the merger, was placed substantially under the control of Raleigh's pre-merger management and based at Raleigh's Nottingham site. This company became responsible for approximately 80 percent of British cycle production.[1] In effect, then, while Raleigh was an unusually foresightful cycle manufacturer until the 1930s, the emergence and subsequent merging of two large conglomerates during the 1940s and the 1950s meant that after this point Raleigh essentially *constituted* the British cycle industry. This situation has only recently begun to change, although Raleigh still maintains the highest market share of all brands in Britain.

Raleigh has always prided itself on being at the forefront of major developments in the cycle industry, at technological, organizational, and commercial levels, and these features too make Raleigh a valuable focus for studying the British industry as a whole. From its earliest days, the company was quick to employ new production techniques to improve quality while cutting costs. This reached a peak between the 1930s and the 1950s with the introduction of Ford-style conveyor systems and substantial modernization of plant, and it culminated in the building of two entirely new factories, one in 1952 and one in 1957.

Raleigh also experienced, however, the industrial strife that was common in British engineering in the early twentieth century. Industrial strife returned with a vengeance in the 1960s and the 1970s. Struggles over wages at Raleigh were exacerbated in the late 1970s by a decline in overseas sales (which in the early 1960s had accounted for some 70 percent of production[2]) and an increase in imports from overseas. Raleigh thus had to deal simultaneously with a changing workforce and a changing world. Until the 1960s, the drive for expansion reflected a commitment to modernization and modernity. In the late twentieth century, Raleigh experienced a struggle with postmodernity and globalization that manifested itself in yet more changes in organization and production techniques and in great reductions in the workforce and in output. While remaining the largest British cycle manufacturer, Raleigh has lost the clearly dominant position it used to enjoy as a result of changes in the wider structure of the industry that are evident elsewhere in manufacturing too. The Raleigh story thus encapsulates a major theme explored in chapter 1: the relationship among technological, social, and cultural change in the shift from modernity to postmodernity. I will trace these struggles at Raleigh, beginning here by outlining how the company established itself within the industry as a prominent innovator of products, processes, and company organization during the late nineteenth century and the early twentieth. I will then look in more depth at the company's history and development, exploring in the Raleigh story the significant themes of modernization, postmodernization, and globalization, labor relations and technological change, and the links between technology and culture in the shaping of a sociotechnical frame.

The Sociotechnical Frame of the Factory Bicycle

The Raleigh company[3] was founded in 1888 when Frank Bowden invested capital in the two-year-old cycle workshop of Woodhead, Angois and Ellis, having made his fortune speculating on property, stocks, and shares while working in Hong Kong. Bowden had returned to Britain because of ill health and had been advised by a doctor in Harrogate to take up cycling. Impressed with his "Raleigh Safety" bicycle on a tour of

Europe, he traced it to its source in Nottingham. Bowden replaced Ellis as the firm's financier, and by the early 1890s he had increased its output from three bicycles a week to sixty, with a parallel increase in the workforce (Bowden 1975).[4]

Raleigh at this time exhibited many of the features of what I will call the sociotechnical frame of the "factory bicycle." This frame emerged during the 1870s as a number of manufacturers and entrepreneurs took what had previously been a hobbyist activity and turned it into an industry. The first of these was the French carriage maker Pierre Michaux,[5] who added pedals to the front wheel of an old Draisienne or hobby horse in the 1860s. The hobby horse had been invented in 1817 by a forest manager in Baden called von Drais. It consisted of two wheels, connected by a wooden frame, with a steering handle on the front wheel. It was propelled forward by straddling the frame and pushing one's feet against the ground. The hobby horse, briefly taken up as a society toy, was improved upon, particularly by the British coachmaker Denis Johnson (Street 1998; Ritchie 1975). In subsequent decades a range of velocipedes (as they became known) were built by engineers from a variety of backgrounds—mostly as a sideline to their main trade (Ritchie 1975). They thus brought a range of skills and techniques into velocipede manufacture.

The innovation of Michaux was not the first time pedals had been added to a velocipede, but it led to the first large-scale production of these machines—known colloquially as "boneshakers"—starting in the mid 1860s. Other manufacturers soon followed suit, refining designs and improving components. Cycling was taken up by the upper classes in Paris and then in London and New York, encouraged by sports competitions that tested the performance of different models and by the appearance of riding schools to help train novices in this initially counterintuitive activity. The proprietor of one riding school, Rowley Turner, was the Paris agent of the Coventry Sewing Machine Company. In 1868, Turner secured orders for his firm to produce 400 velocipedes for the French market, and thus set the British bicycle industry in motion. While French production was curtailed by the Franco-Prussian war of 1870–71, the innovative Coventry Machinists, as they renamed themselves, pushed bicycle production firmly out of the craft workshop and into the factory. A major figure in this transformation was James Starley, described by Ritchie as "probably the most energetic and inventive genius in the history of bicycle technology" (1975: 70). Starley turned the heavy wooden French boneshaker into a much lighter, all-metal machine. The front wheel was made progressively larger to make pedaling more effective, and in the 1870s a range of innovations to wheels, ball bearings, and—in tricycles—the chain and transmission system brought bicycle components close to where they stand today. While Starley and his former Coventry Machinists colleagues George Singer, Thomas Bayliss, and William Hillman became central players in the newly emerging bicycle industry, it was Starley's nephew John Kemp Starley whose low-wheel Rover safety design of 1885, using a chain drive to the rear wheel, quickly became an enduring template for the standard bicycle. This innovation, coupled with John Dunlop's pneumatic tires, led not just to the *closure* and the *stabilization* of bicycle design (Bijker 1995) but also to the development of an obdurate bicycle industry whose training and tooling became increasingly less open to the kind of innovative free-for-all that had characterized it previously (Ritchie 1975; Hudson 1960).

The sociotechnical frame of the factory bicycle is thus characterized by a particular configuration of *people* (innovative engineers such as Michaux and the two Starleys, entrepreneur bicycle promoters such as Turner, racing cyclists and sports promoters, and upper-class leisure cyclists), *technologies* (including the shifting construction of bicycles from wood to iron, the innovations in frame design and componentry, and the factory equipment used in production), and *cultures* (including the innovative artisan cultures in which cycle production was rooted, alongside a fiercely competitive business culture, the sports culture of cycle racing, and the consumer culture of upper-class leisure cycling that quickly spread from the 1860s on).

Raleigh's location within this sociotechnical frame is clear. To begin with, Raleigh was highly innovative both in its products and in its processes. Harrison (1977: 8) cites Woodhead and Angois's "Raleigh Safety" of 1886 as an important development in the progress of safety design, introducing the element of triangulation to the frame, although this is not a feature of the machine depicted in the Raleigh advertisement shown in figure 2.1.[6] Hadland (1987: 12) writes that the first Raleigh models "featured a sprung steering head to absorb road shocks" and "the Raleigh patent interchangeable chainwheel." Such features were tested in races, to which Raleigh's products were subjected from the earliest days of Bowden's involvement, with star riders such as A.A. Zimmerman in the 1890s and Reg Harris and others in the 1940s (Lloyd-Jones and Lewis 2000: 54; Bowden 1975).

The case of Raleigh also illustrates the ease of entry and exit in the industry's early days, as highlighted by Harrison (1977). Paul Angois and R. M. Woodhead, who built the original Raleighs, were machinists

Figure 2.1
An 1887 advertisement for Woodhead, Angois & Ellis. Source: *Nottingham Evening Post*, May 7, 1887.

employed in local engineering factories before they set up their cycle workshop (Harrison 1977: 42). Until he joined them, Frank Bowden had no direct experience of the cycle trade beyond "investigations of the market" made during his convalescence after returning from Hong Kong (Bowden 1975: 16). William Ellis, Woodhead and Angois's original financier, was previously a lace gasser, and he appears to have retained this occupation throughout his time in the cycle industry with Raleigh and then Robin Hood Cycles (White 1894: 491). By 1895 his only entry in *Kelly's Directory of Nottinghamshire* was again as a lace gasser (Kelly & Co. 1895: 281). As Harrison argues, then, it was relatively easy to move into cycle production, and easy again to move on to another trade. Furthermore, Raleigh's occupation of a former lace factory on Russell Street illustrates Harrison's point that cycle production could be adapted to a variety of premises (Harrison 1977: 69).

Harrison describes Bowden as typical of the "financial overlords" who entered the cycle industry at this time, investing their money but leaving the day-to-day running of the factory to others (ibid.: 50). Nevertheless, Bowden made the major decisions affecting the company. He increased Raleigh's share capital in 1889 with a small private incorporation (ibid.: 65) and again in 1891 with a "not too successful" public flotation—one of the earliest in the cycle trade—through which he retained a controlling interest in the firm (ibid.: 53–54, 441; Harrison 1985: 60). This was consolidated in 1896, when a boom in the cycle market over the previous year or so led to a series of flotations in the cycle trade. Raleigh was one of many well-established firms that achieved a full public subscription, managed by the notorious cycle industry speculator and entrepreneur E. Terrah Hooley (Harrison 1977; Lloyd-Jones and Lewis 2000).[7]

Bowden did not just make financial decisions, though. Harrison (1977: 54) describes him as "forceful and growth-minded" and writes that, although he left engineering decisions to Woodhead and Angois, "by 1894, dissatisfied with their policies, Bowden had purchased their interests and virtually pushed them out." The company's committee minutes certainly confirm some antagonism toward the two after their departure,[8] when they were replaced by George Pilkington Mills, a successful amateur racer who had already established several long-distance records in both England and France (Ritchie 1975: 131; Alderson 1972: 160). More importantly, Mills was an example of the new breed of "engineer-managers" that began to emerge in the industry after the boom of 1895–1897, although when it came to making decisions they were usually subordinate to their employers (Harrison 1977: 340). Mills had trained as an engineer at University College, Liverpool, and had worked in marine engineering before joining Dan Albone's Ivel Cycle Works in Biggleswade in 1890. By the time Bowden took him on, he was works manager at Humber & Company's factory at Beeston, Nottinghamshire. Bowden made him chief draftsman at Raleigh (ibid.: 54; Bowden 1975: 16). Bowden charged Mills with designing a new purpose-built factory for Raleigh at Lenton, a few miles from the center of Nottingham, which was opened in 1896 (Bowden 1975: 18). Meeting the cost of opening this new factory was the main motivation behind the public flotation of the same year (Harrison 1977: 459; 1981: 190).

Consolidating the Frame: American Technology and British Innovation

Under Mills's technical control, Raleigh began to take a leading role in introducing American technology into British cycle manufacture, beginning during the boom of the mid 1890s. The collapse of the boom in the United States had led American producers to flood the British market with cheap bicycles. In seeking to compete with these imports, British manufacturers came to recognize the benefits of production methods geared toward higher outputs with lower labor costs. Several companies sent engineers to the United States during this period to study the "American system," including Mills from Raleigh. The result was greater integration of machine tools into British production from the late 1890s on (Harrison 1977: 264). British companies didn't take on these new methods unquestioningly, though, adapting American tools to their own needs and traditions. They sought, for example, machines that would last for several decades rather than simply those that promised a higher output (ibid.: 276). They thus continued to apply a craft-based approach to their new equipment which indicates a failure to recognize that the American system consisted of more than just machine tools (ibid.: 265). Mass production also required, in Harrison's words (ibid.: 331), "a great deal of attention to be given to the surrounding productive apparatus and layout, and to the utilization of labor and raw material inputs."

That said, changes to the organization of British cycle production during this period were substantial. The use of automatic machinery became more widespread, helped by the introduction of electrical power. Chains, chainwheels, and screw threads were standardized. There were also changes to company practices, with less rigorous factory inspection, more active selling techniques, and the introduction of credit purchase systems. New employment structures were adopted, including the introduction of shift work, the employment of cheap female and youth labor, and the dividing up of skilled operations into their unskilled components so workers could then be paid by the piece rather than by the day (Harrison 1969: 298–299).

Many of these changes were evident at Raleigh. The new Lenton factory was equipped with "many new and expensive automatic and other labor saving tools."[9] Mills's trips to the United States in 1897 and 1899 were motivated explicitly by a desire to learn from American methods (Harrison 1969: 295n5).[10] As a result of the earlier trip, Raleigh was a pioneer in bringing sheet-steel stamping into British cycle production. Sheet-steel stamping was used in particular to make the Raleigh tubular fork crown (figure 2.2), a distinctive feature of Raleigh machines that had been introduced in 1892 (Harris and Bowden 1976). This new production method, introduced in 1900, provided Raleigh with one of its early slogans, "The All-Steel Bicycle." Raleigh was similarly a pioneer in the use of liquid brazing for joining frame tubes, another American technique which it adopted in the same year, partially replacing the traditional method of open-hearth brazing. Together, these two innovations reduced labor costs and cut the amount of machining required, reinforcing the factory basis of British cycle production. By the early 1900s,

Figure 2.2
A 1921 Raleigh advertisement showing a tubular fork crown. Source: *CTC Gazette* 30, no. 3.

then, the sociotechnical frame of the factory bicycle was giving greater prominence to production innovations coming from the American industry, although this was only the start of a move toward a British version of mass production. That move would take some decades to become fully established.

While adopting and adapting American technology, Raleigh nevertheless suffered from problems rooted in the industry's craft traditions. Harrison (1977: 289) reports that, after the Lenton Works were completed in 1896–97, Frank Bowden went into semi-retirement "out of an alleged concern for the state of his health," leaving the day-to-day management of the company to D. W. Bassett, who had previously been at Humber's Beeston works with G. P. Mills. Various problems with accommodation, production, and management brought Bowden, in Harrison's words, "scurrying out of retirement to retake the helm" (ibid.: 341). These problems, described by Bowden in his report to the company's 1897 annual general meeting as a "continued run of misfortune,"[11] were due primarily to the slump that affected all cycle manufacturers after the boom; this was exacerbated by overproduction as the anticipated continuing high sales were not realized.[12] On top of this came problems with the new factory, which took longer than expected to fit out, causing damage to stock.[13] Finally, there were conflicts with labor that resulted in strikes and lockouts in the Nottingham cycle trade during 1897 and 1898 (Harrison 1977: 311; Cooper 1993: 10).[14] Raleigh's response was to align itself with the Employers' Federation and to sack workers in order to bring in outsiders.[15] It was at this time that piece rates were first introduced at the factory.[16] This set of events established the company firmly within the difficult labor relations that were already characteristic of British manufacturing and which would go on to shape later sociotechnical frames of the bicycle.

The outcome of Raleigh's problems in 1897–98 was a restructuring of the firm that involved the winding down of the existing Raleigh Cycle Company—which had large debts and no prospect of an increased bank overdraft—and a reconstruction centered on the Gazelle Cycle Company.[17] Gazelle had been established to meet the demand for cheaper bicycles during the recent boom without compromising the reputation of the Raleigh brand.[18] Bowden had been unable to sell Gazelle as he had wished to do during the 1898 slump; the new company was subsequently incorporated around the Gazelle business in 1899.[19] Under Bowden's control, this new version of Raleigh set off down a path of modernization, increased productivity and expansion, focused on "personal capitalism" and on a commitment to "quality" products (Lloyd-Jones and Lewis 2000). This included the establishment of the Sturmey-Archer gear department, which produced hub gear systems based on patents that Raleigh had bought and then developed.

Bowden reinforced his grip on Raleigh after struggles with shareholders in 1903–04 (ibid.) and after a cash crisis in 1908, when the tying up of capital in the new credit payment scheme and poor sales in the previous year (attributable to bad weather) had compounded a large bank overdraft (Bowden 1975: 28–29; Cooper 1993: 11; *Times*,[20] November 13, 1907). The resolution of this crisis was an offer by Bowden to inject extra capital into the company provided all other shareholders agreed to be bought out; this was accepted "with only two dissentients" (*Times*, August 1, 1908), bringing the company back under Bowden's sole control. From this point on, Raleigh pursued a vigorous business strategy that involved substantial investment and a continual expansion of production which pushed it toward a Fordist type of approach that it never actually reached, precisely because of the mismatch between the culture of the British industry and the American meanings associated with the technology being adopted (Lewchuk 1987; Bowden 1975; Cooper 1993).

After the 1908 shakeup, Raleigh entered a period of consistent success through investment and expansion that continued until the 1950s. The factory at Lenton was expanded in 1922, and extensive modernization was carried out throughout the 1920s and the 1930s. A second factory, costing £1.25 million, was built in 1952. With a further extension in 1957, which cost £5 million, the company occupied 64 acres of land, with more than 6 miles of overhead conveyors transporting cycles and parts around the site. The workforce by this point had reached a peak of about 8000. This program of factory expansion coincided with an expansion in the number of brands owned by Raleigh, something that indicates a deliberate strategy pursued over many decades. Having set up Gazelle in the 1890s, Raleigh bought the Robin Hood brand in the early years of the twentieth century with the same objective: to sell increasing number of low-priced bicycles not visibly associated with the Raleigh brand (Bowden 1975).[21] After many years in disuse, Gazelle was re-launched in 1938, then renamed as Robin Hood Cycles in 1943 to avoid confusion with the then-unrelated Dutch Gazelle brand (ibid.: 71, 76).[22] However, before this re-launch Raleigh had already been engaged, along with other large manufacturers, in a "buying spree" (Raleigh Industries Ltd. n.d.) that continued until the 1960s, buying up smaller cycle companies which had met financial difficulties. In 1932, during the world slump that followed the 1929 Wall Street crash (which seems, if anything, to have boosted cycle sales as a relatively cheap means to mobility in times of uncertainty—see Lloyd-Jones and Lewis 2000: 114–115), it bought up Humber Cycles. In 1935 a takeover of Hercules, Raleigh's major rival by this time, failed at the last minute because Hercules changed its terms for the planned merger. In 1943, though, Raleigh bought Rudge-Whitworth. It bought the cycle interests of Triumph and Three Spires in 1954, and those of BSA in 1957.

In 1960, Raleigh made its final large-scale merger, with Tube Investments (TI), and became that company's Cycle Division. It absorbed in the process its last major competitor, the British Cycle Corporation, which had been formed when TI bought Hercules some time after the Raleigh-Hercules merger had fallen through. In the same year, Raleigh bought Carlton, a small hand-builder of racing cycles. By this point Raleigh was responsible for more than 80 percent of British cycle production. The last notable cycle brand Raleigh bought before the 1980s was Moulton Cycles in 1967, some nine years after dismissing Alex Moulton's prototype of a small-wheel suspension bicycle as "an overgrown 'fairy cycle'" (Roy 1983: 73).

Raleigh's phenomenal success during the first half of the twentieth century was followed in the 1960s and the 1970s by a severe decline that matches closely the patterns of industrial change cited in debates about "Fordism" and "post-Fordism" (Gilbert et al. 1992; Harvey 1989). In chapter 3 I will begin to trace these developments by focusing on the program of modernization in the cycle industry that complemented developments in transportation and leisure culture in the formation of a *sociotechnical frame of the mass bicycle.* This frame brought together a system of labor relations within the industry which had its roots in craft production, American-inspired production technology, a transportation culture geared toward mass cycle commuting, and a leisure culture that emphasized cycling as a means of escaping the city.

3

Modernization, Competition, and Collaboration in the Sociotechnical Frame of the Mass Bicycle

Heterogeneous Engineering in the Works: The Roles of Sir Harold Bowden and William Raven in the Modernization of Raleigh

An important element in Bijker's (1995) exposition of SCOT is the need to set the individual biographies of "great inventors" within a social context. Individual life histories can, though, benefit our understanding of sociotechnical change by pinpointing the contributions that individuals make toward how a sociotechnical frame becomes transformed over time. Like Bijker, I am concerned here with the interplay between structure and agency, and between continuity and change, in the shaping of a new sociotechnical frame. In particular, I will concentrate in this chapter on changes in the organization and technologies of production at Raleigh during the interwar years, and on how these changes were shaped by the relationship between two very different characters within the company.

From 1894 to 1959, fortnightly meetings were held by the Raleigh "committee," which generally comprised the Managing Director, the Chairman, the Works Director, and the Works Manager, along with other Directors and senior works staff at various times. A striking feature of the minutes of these meetings is the contrast between two major figures in the company's history during the 1920s and the 1930s: Sir Harold Bowden, the son of company founder Sir Frank Bowden, and William Henry Raven, the Works Manager and later the Works Director (figures 3.1 and 3.2). Though I will not attempt any kind of biography of either man, it is worth noting some of the main features of each of their lives as they appear in Raleigh documents.

Harold Bowden joined Raleigh around 1900 when Sir Frank asked him to leave Cambridge University prematurely to help run the company during its post-boom difficulties (Bowden 1975: 27). By the end of

Figure 3.1
Sir Harold Bowden. Source: *Bicycling News,* 1921. Reproduced from Bowden 1975.

Figure 3.2
William Raven. Source: Raleigh Industries 1952b.

World War I, Harold was virtually running the company due to his father's increasing age and ill health. In 1921 Sir Frank died. Harold inherited both his father's baronetcy and his role as Managing Director and Chairman of Raleigh.

Sir Harold Bowden was a very public figure both locally and nationally. In addition to local "good works," he was a member of a government committee concerned with education in the late 1920s.[1] In 1931 he was elected Chairman of the British Olympic Association (*CTC Gazette* 40, 1931, no. 5: 151). He published articles on education, sport, and many other topics during this period. He also wrote frequent articles for national and local newspapers on the state of the British economy and British industry, dealing with issues such as labor relations, the production methods of British industry, thrift and efficiency, and free trade. In 1928 he joined a conference of major industrialists, government, and trade union representatives aimed at resolving the industrial strife of the previous few years,[2] to which his major contribution was a twelve-point "Industrial Code" that drew on his own proclaimed approach to industrial relations—the principal feature of which was open access to management for any shop-floor worker with a grievance (*Westminster Gazette*, January 12, 1928).[3]

Bowden claimed "that he had never taken part in the war which some people said Capital was eternally waging against labor."[4] He was nevertheless antagonistic toward organized labor and the political left. One of the most visible examples of this is an argument over the Conservative government's industrial policies during the late 1930s in the letters pages of the *Nottingham Journal* in which Bowden attacked "communists and socialists" who he believed were envious of the success of capitalists at running their businesses.[5] His approach to industrial relations has been termed "personal capitalism" by Lloyd-Jones and Lewis (2000); it is perhaps more accurately described as a kind of welfarist paternalism reminiscent of what Smith et al. (1990) term "Cadburyism," referring to the benevolent management style of the Quaker-owned Birmingham chocolate firm.

Bowden presided over Raleigh as Managing Director until 1938, when he retired from this post. He remained Chairman until the 1960 merger with TI, taking a less active but still important role in the company's affairs. He died around the time of the merger's completion.

In contrast to Bowden's prominence in public life, William Raven is all but invisible outside the Raleigh committee minutes. Discovering much detail about Raven's career is difficult. The most interesting source is Tony Hadland's *Sturmey-Archer Story* (1987), which includes quotations of personal attacks on Raven from a colleague at Raleigh, I. C. Cohen. Raven replaced G. P. Mills as works manager in 1907.[6] According to Hadland, he was made responsible in 1908 for overseeing production of Sturmey-Archer hub gears (ibid.: 50). In 1909, according to Cohen (quoted in ibid.: 57), "Raven imported all his pals from Birmingham and planted them in strategic positions throughout the works. The Gestapo hadn't much they could teach him . . . and even when I left the firm for the first time Raven wanted to search my belongings." Cohen's hostility toward Raven appears to derive largely from a conflict over the authorship of a motorcycle hub gear patent which was registered in the name of Raven and Frank Bowden (ibid.: 55).

The next development in the Raven story, according to Cohen, is that he left Raleigh in 1918 "in a great hurry," taking with him most of Raleigh's senior sales staff in order to set up a rival company (ibid.: 60, quotation from Cohen). Cohen's claims must be treated with some caution, since Raleigh's committee minutes show that Raven was still an active Works Manager up to and including the meeting of 12 January 1922. By the next meeting, though, on 1 February, Raven *had* left the company, and the upheaval Cohen indicates is confirmed. The minutes state that Sir Harold Bowden had been to visit each foreman individually to ascertain the amount of support for Raven's replacement, A. R. Holland, who 2 years earlier had come to Raleigh from BSA.[7] Sir Harold found "each of the foremen professing loyalty to the Company and expressing an intention to give Mr. Holland all possible support."[8] No further explanation is given in the minutes.

What is most surprising after Raven's apparently problematic departure is that he rejoined the company in 1929. Leonard Clarkson, in an interview for the Nottinghamshire Oral History Project of 1982–84, tells how he joined the Ray Cycle Company in 1925, working for Raven there until 1929, when Raven was asked to become Works Manager again at Raleigh. Clarkson was left to wind down Raven's company, and then joined him at Raleigh in the drawing office. He eventually replaced Raven as Works Manager when the latter was promoted to Works Director in 1937. Clarkson became Works Director himself when Raven retired in 1942 at the age of 72.[9] In accordance with Clarkson's account, Raven reappears in the committee minutes in October 1929 with no undue attention, taking on the same role in the committee that he had had until 1922. It isn't until December, though, that Raven returns to the post of Works Manager.[10]

Apart from the comments in Hadland's book, Raven is barely mentioned in published accounts of Raleigh. In 1938, when Sir Harold Bowden retired as Managing Director, reports mentioned that "Billy Raven" presented Sir Harold with a gift from the works staff, and that Sir Harold "spoke in eulogistic terms of the three new directors, and of Mr. Raven, who had brought the works organization to the perfection it was today."[11] A similar recognition of Raven's contribution to Raleigh appears in the souvenir brochure of the opening of the new factory in 1952. This counts Raven's retirement 10 years earlier, after 45 years' service, as one of the "milestones in Raleigh's history," ranking in significance alongside the founding of the company, Sir Harold Bowden's joining the Board, and previous factory expansions (Raleigh Industries 1952).[12] In the same year, a brochure commemorating the fiftieth anniversary of Sturmey-Archer included a few lines about Raven (Raleigh Industries 1952b).[13] A 43-page unpublished history of the company written by Sir Harold in the late 1940s also pays tribute to Raven, stating that he "was largely responsible for the high standard of Raleigh quality and the works organization, and kept the factory equipped with all the latest machinery that was devised."[14] Bowden also refers to Raven's receipt of an OBE in the New Year's Honours List for 1942.[15] Surprisingly, then, there is no reference to Raven in Gregory Bowden's 1975 history of the company or any other account of Raleigh's development that I have seen, and this reflects the absence of production from much of cycle history (Millward 1995; exceptions are Grew 1921; Hudson 1960; Harrison 1977). From the evidence of the committee meeting minutes, though, it was primarily Raven, with the cautious support of Sir Harold Bowden, who brought about the massive expansion of Raleigh's manufacturing capacity during the 1920s and the 1930s—a role for which the Raleigh Board of Directors agreed to pay him £2500 in 1944 after he had retired, in order to prevent him working for anybody else. This was considered "essential for the protection of the Company's trade."[16]

The Heterogeneity of Cycle Production

Raleigh's fortnightly committee meetings from 1894 through 1959 show that the company was continually dealing with modernization, new technology, and changing organizational practices in order to keep up with competitors, the market, and a changing world. Raleigh was, in other words, engaged with the question of whether and how to move beyond the sociotechnical frame of the late-nineteenth-century and early-twentieth-century *factory bicycle*, whose production had begun to require the use of American equipment but was still rooted in the practices and products of its craft-based roots. As leading members of the committee, Bowden and Raven were striving in this period to find a balance between several disparate elements, acting as what John Law (1987) characterizes as "heterogeneous engineers." These elements included not just the commercial outputs of the company but also the more mundane aspects of managing a complex organization (table 3.1). Most obviously, there were the company's products, which varied over time as new lines were introduced or old ones were discontinued. Then there were the raw materials for making them, which necessitated consideration of their availability and, above all, their cost. New plant for the factory came under discussion as Raven and others investigated and acted upon options for new purchases, as did production processes, which the committee would assess and then perhaps install. The factory buildings themselves required maintenance, repair, and extension. Alongside all these were the various sections of the company's workforce, with particular attention paid to their wages and levels of productivity. Running throughout discussions

Table 3.1
Elements that the Raleigh committee had to manage.

Products
bicycles, gears, brakes
at times: automobiles, motorcycles, motorcycle parts, munitions
Plant
production processes
raw materials
factory buildings
Workforce
skilled and unskilled workers
foremen, charge hands
trade union shop stewards
women and youths employed in the works
returning World War I soldiers
various categories of office staff rarely mentioned in the minutes
Financial considerations
wage costs
costs of plant and materials
productivity
External factors
other cycle firms
supplier industries
fellow employers
Raleigh's distributors
consumers
the weather

of all these elements was the question of cost, with production costs balanced against revenue.

Costs were also important in considerations of external factors. These arose most notably in relation to other firms, whether as suppliers of raw materials (notably steel and coal), as competitors whose model ranges and prices Raleigh needed to keep in mind, or, during periods of industrial tension, as fellow employers and owners of capital whose common class interests with Raleigh had to be balanced against the committee's desire not to have production affected by strikes in its own works. Keeping aware of Raleigh's position on the scale of wages and conditions relative to other employers was important for the committee. Other external elements that affected the committee were feedback from distributors on consumer demand and on responses to particular models and the cycle industry's biggest historic enemy: the weather. It is notable that little direct mention is made in the minutes of consumers themselves; presumably, questions of product design and marketing were dealt with outside the committee.

The minutes chart the tension between expansion and expense that framed all activity at Raleigh. This tension is expressed most clearly in the contrast between the modernizer Raven, who was charged with implementing many of the committee's decisions, and the financially cautious Bowden, who often sought to rein in Raven's more expensive ambitions.

Modernization and Efficiency

Harold Bowden, like his father, was committed to expanding his company. He was willing to invest huge sums in new equipment to modernize the factory. At the same time, though, he was very careful with his money, whether he was dealing with small expenditures or with plant costing thousands of pounds. In 1927 he wrote an article espousing the value of thrift and efficiency and advocating a "nation of Micawbers." He wrote that "in America they worship the god of efficiency with a logical thoroughness which is to be wholly recommended."[17] As I. C. Cohen claimed, "the Bowdens, father Frank and son Harold, were "eagle-eyed on every fraction of a penny" (quoted in Hadland 1987: 35). The committee minutes consequently feature regular observations from Bowden concerning the economic efficiency of the factory. For example, the following entry appears in the minutes for November 9, 1921: "Sir Harold mentioned that in searching for certain information that he required for a purpose outside the scope of the Raleigh business, he had formed the opinion that the present efficiency of our workmen as compared with pre-war stood at approximately 75 percent, while on the other hand, the present cost of production, both in bicycles and gears was 2½ times the amount paid in 1913. He suggested that greater efforts should be made to remedy these matters."[18] Three months later Bowden went on a walk around the factory and found "a distinct improvement in the general morale of all employees . . . The old tone of indifference was less apparent and everybody seemed to be taking a keener interest in their work."[19] This concern of Bowden's relates back to problems the company was experiencing earlier in the same year. That January, during a period of slack trade, he had pointed out that Raleigh's financial situation was being adversely affected by high payments for wages and materials. He suggested then that "only by rigid economy in every direction could [the company] weather the storm of depression," and recommended measures to achieve this: cutting all unnecessary purchases, buying necessary materials only in small quantities, and cutting wages, especially those of "unproductive" (i.e., office-based) employees.[20]

The response to this proposal is interesting in light of Harrison's comments about early cycle companies that were financed by entrepreneurs with little knowledge of the actual running of the company, which they left to engineer-managers such as G. P. Mills and William Raven (Harrison 1977: 50, 340–341). Both Frank and Harold Bowden were unusual as "financial overlords" in that they did take a close interest in the day-to-day running of Raleigh, and Harold in particular attended committee meetings as a matter of course. It is a little surprising, then, to find in his comments of January 1921 a lack of understanding of the workings of the shop floor, which Raven was able to challenge and thus use as a means of tempering Bowden's proposals. In response to being called upon "to cut in every possible direction," Raven pointed out the need to maintain the loyalty of foremen and other leading staff. He consequently recommended not cutting wages but instead encouraging foremen to "take a more active part in their respective departments" in order to enable a reduction in the number of charge hands and stores assistants.[21]

Similar exchanges occur elsewhere in the minutes, with Raven frequently being called upon to justify what to Bowden seemed unnecessary expenditure. The usual response to Raven's explanations is a call for increased production.[22] These interactions between the two men illustrate the tension between Raven's practical understanding of the shop floor and of the requirements of increased production and Bowden's concern to obtain the best possible value for his money. It also indicates the huge investment costs associated with Raven's Ford-like agenda in keeping the factory well prepared for long production runs.

Raven wasn't just manager of the works *staff.* It was his responsibility also to carry out Raleigh's postwar "reconstruction" program (completed around the summer of 1921) and the subsequent massive factory modernization (which spanned much of the 1920s and the 1930s, partly under Holland's management during Raven's absence). Yet this role too was subjected to Bowden's price-conscious scrutiny, hinting sometimes at a degree of personal tension between the two men. Even as the modernization program was nearing completion in February 1935, Bowden continued to require Raven's assurance at committee meetings that orders for new plant came within the company's overall program.[23] Raven's responses to pressure on spending from Bowden show a careful balance between compliance with the need for cost cutting and persuasion that expenditure was necessary for modernization. Ultimately, though, it was Raven who had to give way. When he tried to suggest in March 1935, while Bowden was away on a world tour, that new tools could be bought without permission from the committee, he was contradicted by others on the ground that Sir Harold had instructed that tools needed to be sanctioned along with plant.[24] Later in the same year Raven was again under pressure to restrict his spending on new plant.[25]

Despite Raven's commitment to modernizing the factory, his wishes were always subordinate to those of his boss, which were rooted in a concern with costs. The various elements that Raven (and the committee generally) had to juggle in the course of heterogeneous engineering were not, then, undifferentiated; some elements, including technical constraints, the weather, and especially Bowden's financial caution, were less malleable or open to negotiation than others. Nevertheless, Raven managed to ensure that Raleigh's factory was by the mid 1930s among the most up-to-date in Britain.

Competition and Innovation in the Modernization of the Industry

I want in the following pages to trace Raleigh's interwar expansion program in some depth, first in relation to arguments about the degree to which British engineering firms adopted Fordist mass production and then in order to set these changes within a SCOT framework. A number of related themes in the committee minutes are pertinent to these questions:

- a concern to increase production figures and to monitor the production levels and model ranges of competitors
- attempts to improve Raleigh's own range and to produce it more cheaply
- a pursuit of new production processes along with the necessary new plant and components that could enable the production of better quality but cheaper products.

These factors embody some of the principles of the American system and Fordism, but they come into conflict with others.

Products and Prices

As I mentioned above, Sir Harold Bowden was greatly impressed with the American quest for efficiency. During the 1920s and the 1930s, Raleigh—

often drawing directly on American methods—adopted a variety of practices that aimed to emulate this quest. Bowden's primary concern was presumably to beat the competition, and this is manifested explicitly in the committee minutes in a regular monitoring of differences between Raleigh and other companies in terms of models available, prices of cycles and parts, and (occasionally) wages and piece rates.[26]

At Raleigh and at other companies where quality was emphasized more than prices (Harrison 1977; Lloyd-Jones and Lewis 2000), price cutting was a serious problem. An illustration of the effects this could have came in March 1921 when Rudge-Whitworth made a "sensational cut in prices." The Raleigh committee was aware that "sooner or later we should be compelled to do something similar,"[27] although Harold Bowden achieved an agreement with three other cycle firms not to follow suit.[28] Nevertheless, Rudge-Whitworth's cuts depressed sales and caused the Raleigh committee to anticipate cuts in production. Consequently, the committee adopted a number of strategies to deal with the various effects of price cutting. In particular, it cut the size of the workforce and put pressure on remaining workers to accept lower wages and piece rates. These actions were compounded by the fact that the works was by now on short time. By autumn it was down to a three-day week by the autumn, and production was down to between a half and a quarter of what it had been.[29] This caused output to decrease almost 50 percent in a year (table 3.2).

Another factor in price cutting was the problem of cheap imports.[30] Gregory Bowden writes that his grandfather, Sir Harold, bought one of the better German machines, presented it to his works staff, and asked them whether a British firm could compete with it on price. Sir Harold is quoted as having said: "Like sensible men they agreed it could not. Our cheapest price for a bicycle was £14, the German's was less than half that sum." (Bowden 1975: 47) Gregory Bowden (ibid.) notes that "the result of that most significant meeting was that the workers agreed to take smaller wages and to increase their output in terms of quality and quantity substantially, in return for greater security of employment." While the complicity of the workforce implied here deserves to be questioned, this arrangement indicates that Raleigh's competitive strategy included some elements of the Fordist approach, offering job security in exchange for greater productivity, despite the absence of the ultimate Fordist exchange of high wages for control of the shop floor (Lewchuk 1987; Hounshell 1984).

The most direct strategy Raleigh could adopt to beat its competitors was to increase its own model range and cut prices. The minutes from

Table 3.2
Bicycles produced by Raleigh per year, 1896–1960. Source: typed sheet, NA DD 1267/1.

1896–97	7,813
1897–98	6,068
1898–99	8,051
1889–1900	11,120
1900–01	12,190
1901–02	12,137
1902–03	13,475
1903–04	9,719
1904–05	16,555
1905–06	28,156
1907–08	32,577
1908–09	33,431
1909–10	36,588
1910–11	43,079
1911–12	46,945
1912–13	52,136
1913–14	55,022
1914–15	34,347
1915–16	28,000
1916–17	23,939
1917–18	23,336
1918–19	29,205
1919–20	62,824
1920–21	33,789
1921–22	50,156
1922–23	70,737
1923–24	90,281
1924–25	98,612
1925–26	120,190
1926–27	96,920
1927–28	114,072
1928–29	106,315
1929–30	110,074
1930–31	89,175
1931–32	167,615
1932–33	225,008
1933–34	294,601
1934–35	355,210
1935–36	382,828
1936–37	473,056

Table 3.2
(continued)

1937–38	393,766
1938–39	411,365
1939–40	356,387
1940–41	180,426
1941–42	152,296
1942–43	187,591
1943–44	192,742
1944–45	179,144
1945–46	363,044
1946–47	528,430
1947–48	594,387
1948–49	705,486
1949–50	820,388
1950–51	958,199
1951–52	934,313
1952–53	969,331
1953–54	1,081,823
1954–55	1,047,846
1955–56	1,301,362
1956–57	1,118,921
1957–58	1,146,871
1959–60	1,197,585

1919 include a comment that Raleigh would not be manufacturing the cheaper Robin Hood machines in the 1919–20 season,[31] indicating that the production of cheap bicycles under a different brand name was considered either no longer viable or no longer necessary. During the 1920s, as Raleigh consolidated its position after the postwar reconstruction program, an alternative strategy was pursued. Rather than introduce a new brand (aside from the brands acquired through expansion), Raleigh expanded its range. This was accompanied by a downward trend in prices and a broadening of the price range across the brand. Old models that had not been produced during the war were reintroduced, and new models were developed. Especially during the reconstruction of the early 1920s, the Raleigh brand was built back up to its prewar state, while prices were regularly reduced. The first model to be reintroduced was an "all-weather" bicycle, first proposed to the committee in March 1920. In June, delivery models were reintroduced to take up the slack caused by a general drop in demand for cycles. In July, proposals were made to produce "a real juvenile machine," and in December it was decided to reintroduce the renowned cross-frame Modèle Superbe (figure 3.3) in response to demand from distributors.[32] These models, reintroduced mainly in response to everyday changes in demand, gradually became visible in Raleigh's brochures alongside newer racing and roadster bikes.[33] By the late 1920s the company's range and prices were very close to those of 1913,[34] a result both of the changing economic climate and, more importantly, of the modernization of the company's production capacity. This was achieved without appearing to compromise either quality or the company's reputation. A 1931 magazine article on the Raleigh factory notes that serious cyclists were impressed at how "the wonders of production . . . have made possible a perfect travel instrument selling at so very moderate a figure" (*CTC Gazette* 40, 1931, no. 5: 151).

The requirements of mass production meant that large manufacturers in fact produced only four types of bicycle at this time: the roadster, the tourist, the sports, and the semi-sports (Hudson 1960: 145). The subtle differences within Raleigh's and other companies' model ranges indicate, therefore, Sloanist tendencies toward flexible mass production—tendencies in which styling, finish, and accessories were central (Hounshell 1984). Nevertheless, Raleigh was able to use its new production equipment in ways that retained credibility while reducing prices, and the strength of the model range was a crucial element in its ability to compete successfully.

Industrial Collaboration in the Innovation Process

Modernization was by no means unique to Raleigh at this time. The cycle industry as a whole (and indeed British manufacturing generally) adopted a great deal of new production technology during the interwar years, and this had significant impacts on the industry's structure—impacts that highlighted tensions in the culture of cycling. Hudson (1960: 131) describes how the industry became increasingly divided during this time between large-scale mass producers and smaller custom builders and assemblers and connects this division with a new tendency after World War I toward amalgamation and hence concentration of production into fewer and larger units. By 1935, nine companies, each employing 500 or more people, controlled 75 percent of the industry's gross output (including parts and accessories) and 83 percent of complete bicycles and tricycles (ibid.: 129). Of these nine, Raleigh, BSA, and Hercules accounted for 55 percent of complete cycles and 70 percent of total employment in the trade. Hercules was the largest of the three, producing 600,000 cycles in 1935—twice as many as Raleigh (ibid.: 130).

SPECIFICATION :

For terms of guarantee see page 24

FRAMES. 24 in. and 26 in. Centre of crank bracket to ground 10½ in.
WHEELS. 26 in. × 1½ in. 28 in. optional.
TYRES. Fort Dunlop, on Roman aluminium rims.
GEAR. Sturmey-Archer 3-speed, 52 in., 69 in., 93 in., or as ordered.
GEARCASE. Raleigh patent detachable oil bath.
SADDLE. Brooks' Supple Top.

PEDALS. 4 in. rubber.
CHAIN. Coventry Elite, ½ in.
STEERING LOCK. Friction band.
FITTINGS. Tools, toolbag and inflator complete.
REFLECTOR. Fitted to rear mudguard.
LAMP. Miller's electric.
BELL.
FINISH. Green enamel, 22 ct. gold leaf lining.

A Lady's Model Superbe is also supplied at the same price.
Finish in black enamel or all-black without extra charge.

Price : £15 · 0 · 0 Net Cash

or 12 monthly payments of £1 · 8 · 0

For full particulars see Order Form, page 21

Figure 3.3
The gent's Model Superbe. Source: 1929 Raleigh catalog.

Hudson uses Hercules as an exemplar of how "growth and prosperity could be attained in this period" (ibid.: 135), and Hercules provides an interesting contrast with Raleigh despite their shared place at the forefront of British cycle production.[35] Hercules was founded in 1910 as an assembler, but by the mid 1930s it had become "by far the largest manufacturer in the UK," producing many of its own components as a means of gaining independence from outside suppliers. It achieved this expansion by increasing its sales to working people through the introduction of no-deposit, low-payment credit purchase, accompanied by large-scale advertising and the sponsoring of cycle racing and trials (ibid.: 137–138). Hercules thus epitomized for the British cycle industry the adoption of Ford's practice of integrating in every possible direction (cf. Hounshell 1984).

There are two significant caveats to this. In 1946, Hercules was bought by Tube Investments, and this gave TI the edge over Raleigh as the largest cycle manufacturer in the world (Hudson 1960: 176). The cycle and cycle-related companies that constituted TI's new Cycle Division retained their separate factories in competition with one another (ibid.). In contrast, Raleigh's policy of multi-directional integration bore a closer relation to the strategies of American companies such as Pope (the firm that established American cycle production in the late nineteenth century) and Ford (in its later Model T "high Rouge" era) (Hounshell 1984; Norcliffe 1997). Raleigh put new brand acquisitions such as Humber and Rudge-Whitworth into production in its own factory, thus gaining not just a new brand but also a new network of agents linked to the new brand (Bowden 1975).

Hercules also diverged from the Fordist model in making no attempt to develop product innovations alongside its innovations in production. It "refused to prejudice its success by chasing eccentricities that appealed only to the few," choosing instead to wait until "new developments had proved their profitability." Then "Hercules brought them quickly within reach of the ordinary cyclist" through mass production (Hudson 1960: 138). This unusual feature of Hercules's approach underlines Hudson's view that while there was intense competition between the larger producers, this did not extend to their relations with smaller assemblers. Rather, there was complementarity between the two sections of the industry. Large manufacturers were unable to cater to the specialist demands that were the main source of custom for the smaller producers and were therefore not a threat to them. At the same time, though, as the Hercules case shows, the smaller provided a continual supply of innovations for the mass producers to draw on once they had been tested in the specialist field (ibid.: 143). This situation echoes, in part, both the claims of Piore and Sabel that mass production and flexible specialization have always coexisted and the criticisms of their position that question their rigid distinction between the two forms (Piore and Sabel 1984; Sabel and Zeitlin 1985; Williams et al. 1987). The British cycle industry of the interwar period used two interdependent production approaches that, as Hudson makes clear, could not have supplied all the needs of the cycle market individually. The "choice" that Piore and Sabel present as having been made at the "first industrial divide" in the early twentieth century in favor of mass production included of necessity a simultaneous commitment to smaller-scale production methods both to support mass production (e.g. in the manufacture of machine tools) and to provide goods for niche markets that mass production could not supply.

The fact that Hercules could choose not to develop product innovations, in the knowledge that it could benefit from the innovations of others, suggests that the relations between and within social groups in the production sphere were an important factor in the British sociotechnical frame of the *mass bicycle* that was beginning to become established at this time. Such connections between companies, which were at least as significant as their rivalries, are missed by many accounts of industrial change based on analyses of American automobile production, despite the free flow of staff among firms even in the mid twentieth century. As Hounshell (1984) shows, the spread of ideas among industries and individual companies—both by the mobility of staff and by Ford's practice of courting publicity—was crucial to industrial progression from the American armory system to flexible mass production.

Another way of understanding the growing homogenization of the cycle industry—at least of its larger-scale component—lies in the notion of *institutional isomorphism*, defined by DiMaggio and Powell (1983) as the homogenization that often occurs within an "organizational field" (essentially equivalent to an industry or a sector) as a result of three possible mechanisms: coercive isomorphism, where organizations are obliged to institute changes because of political influences such as legislation; mimetic isomorphism, where organizations respond to commercial and organizational uncertainty by using the activities of others within the same field as models for change; and normative isomorphism, where professionalization within a field results in similar practices being adopted across a range of organizations. In all cases, different organizations within a field come closer together in their ways of organizing. This analysis is valuable in identifying mechanisms of organizational and industrial change. For example, the concept of normative isomorphism, according to which the professionalization of staff across an industry leads to change becoming structured in similar ways among different organizations, provides a framework against which to set Hounshell's (1984) insights into how new production methods were transferred across a range of industries via the fluidity of personnel. The process of professionalization can certainly be seen at play in the development of the early bicycle industry—several sources (the Raleigh minutes; Harrison 1977; Bowden 1975) indicate links between professional engineers and change across the industry during the late nineteenth century and the early twentieth century. William Raven's career perhaps epitomizes the ability of senior works staff to move easily between cycle companies—initially from Hudson to Raleigh, then from Raleigh to his own company, Ray, and then back to Raleigh again. It might be valuable to take this further by tracing how such processes differ between the formal professionalization referred to by DiMaggio and Powell and the earlier, emergent professionalization that was developing alongside the new industries Hounshell discusses.

As the twentieth century progressed, it seems, the primary cause of homogenization within the cycle industry came to be mimetic rather than normative isomorphism. This is made clear by the evidence in the Raleigh minutes of collaboration among cycle firms on the adoption of new production technology, and it is supported by the accounts of Hudson and others. In contrast to the developments in the American auto industry (led by Fordist standardization and Sloanist flexible mass production), the progress of British cycle production after World War I was not the work of just one or two companies. Rather, the tendency toward an ever larger scale of standardization and mass production was supported by a great deal of sharing of ideas and experiences among companies, something that occurred alongside fierce competition at a commercial level. The development of a new sociotechnical frame of the *mass bicycle* in Britain was thus a collaborative venture among social actors that shared common goals as well as common markets.

It is not surprising, then, that Hercules, Raleigh's major competitor, is mentioned frequently in Raleigh's committee minutes, both as a competitor and as a model of industrial practice. During the late 1920s and the 1930s, the committee clearly regarded Hercules as the company it most needed to watch. In September 1929, Raleigh hired an analytical chemist to test and compare Raleigh and Hercules frames.[36] In 1935, Raleigh attempted to buy Hercules, but the merger fell through when Sir

Edmund Crane, Hercules's owner, changed his terms at the last minute.[37] This led to serious concerns about how to compete with Hercules on price[38]—in January 1936, for example, the committee was gratified to see that Raleigh's sales had begun to overtake those of its rival during the previous year. The competition between Raleigh and Hercules continued until Raleigh's 1960 merger with Hercules's parent company, Tube Investments. From my conversations in the early 1990s with Harold Bowden's son Frank, it appears that relations between Raleigh and Hercules had become less than friendly by that time.

Despite the competition for sales, there was a great deal of cooperation among cycle companies in the sharing of new ideas and methods regarding factory organization and modernization. This issue is closely related to the question of how much the British cycle industry adopted the practices and production methods of Americanism and Fordism. Raleigh was typical of British engineering firms in its incomplete adoption of American methods: it adopted American technologies but not the relations of production that transferred control over the shop floor from labor to management in companies such as Ford (Lewchuk 1987). It is nevertheless worth noting the degree to which Raleigh *did* adopt elements of the Fordist agenda. Many of the technological and organizational concerns of the American system and of Fordism were crucial also to the shaping of Raleigh's operations. Raleigh was constantly striving through the 1920s and the 1930s to eliminate inefficiency and to improve the productivity of its factories, and this was where William Raven was most important to the company.

Learning new techniques from other companies was a practice used at Raleigh as far back as G. P. Mills's visits to the United States in the late 1890s. However, it was under Raven's management of the works that this became most prominent, and the minutes show that this kind of informal contact and sharing of ideas and experiences among manufacturers was a crucial part of the culture of the cycle industry at this time, a factor that greatly facilitated the shift from one sociotechnical frame to another. Raleigh frequently consulted other companies before buying new equipment. In July 1919, Raven viewed BSA's spraying plant before buying new plant himself.[39] In January 1924 the company was planning to manufacture its own rims and visited the factories of two other companies to study their equipment, although these plans were hindered and curtailed over the following few months through efforts made by Dunlop to keep Raleigh as a customer rather than have it as a competitor.[40] In October 1931, Raven recommended installing a large amount of conveying equipment. The committee asked George Wilson to follow this up by contacting the firm that had supplied conveyors to Hercules.[41] The following July, Raleigh again bought new tube-making plant from Hercules's supplier.[42] It appears, then, that Raleigh was taking its lead in modernization in part from Hercules, despite the latter's eschewing of *product* innovation.

Along with informal contact in the course of its factory modernization, Raleigh embarked on more formal arrangements, such as licensing agreements. These were common with early product-based patents, but they also covered production processes. One significant example is Raleigh's adoption of a new brazing process (referred to by Gregory Bowden as "safety brazing") between 1927 and 1929. Bowden describes the process as follows (1975: 53): "The object of the method was to localize the heat generated in the brazing process to the lugs themselves, thus avoiding the risk of metallurgical change in the structure of the tubes with consequent risk of breakage. In order to accomplish this, the Raleigh engineers hit upon the idea of introducing their brazing materials from the inside of the lug and then applying low heat to the outside. It was an extremely simple idea but an effective one."

Safety brazing was not Raleigh's own innovation, though. In September 1927 the company negotiated a sole license for the new process with its inventors, the Victoria-Werke in Nürnberg. Raleigh sent engineers to Nürnberg to "receive the necessary instructions," and in the next few months Raleigh set about installing plant and experimenting with the new technique.[43] By the following March, Mr. Holland, the Works Manager during Raven's absence, reported that "the new system of brazing was much more satisfactory than the old method, forks which had been treated by the new system taking twice the amount of force to deflect compared with forks treated under the old method."[44] The company's 1929 catalog consequently made much of the strength of machines made using this process (figure 3.4).

By the end of 1929 it was evident that safety brazing was not fully compatible with the earlier innovation of sheet steel stamping introduced after G. P. Mills's visit to the United States in the 1890s. In October, at his first committee meeting since returning to the company, Raven "complained of faulty lugs . . . and also of defective low pressure brazing."[45] After consultation with Victoria-Werke, it was found that the new brazing process was less successful when used with pressed lugs (the sleeves which held different frame tubes together) than with machined ones. This problematic encounter between two elements of the sociotechnical

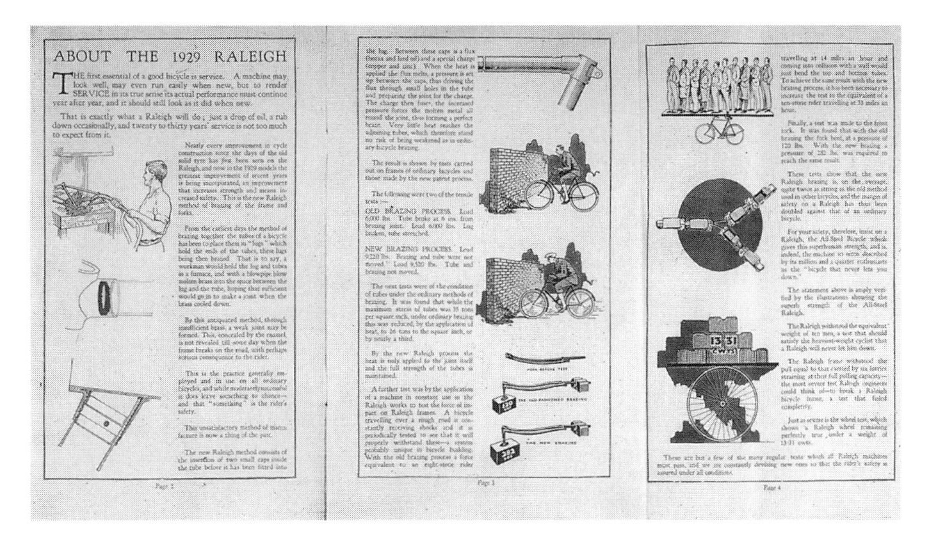

Figure 3.4
"Safety brazing." Source: 1929 Raleigh catalog.

frame helped to further consolidate it by establishing which process was more central to the frame. Raleigh clearly deemed sheet steel stamping more important than safety brazing, since the company reverted to liquid brazing on head lugs and the bottom bracket,[46] the frame joints which are subjected to most stress. In other words, it was the newer of the two processes that was compromised, not the older one. No attempt was made to revert to machined lugs, which would have indicated a step back toward the European factory system and an abandoning of Raleigh's "All-Steel" identity.

Raleigh and Fordism: Vertical Integration and Conveyor Systems

Safety brazing was probably only a short-lived issue for the committee at the time, but it points to a major respect in which Raleigh's approach diverged from that of Ford. Raleigh was always loath to discard old machinery and equipment. In the committee minutes there are occasional references to attempts to sell off old equipment that had been replaced by new plant. I was told by a recent union official that even in the 1990s there were some presses labeled "Property of the War Department." In contrast, Henry Ford himself linked his success with "the economy of scrapping the old equipment immediately upon the invention of the better equipment" (1926: 820).

Another important element of Raleigh's modernization, and one that again drew on information sharing, concerned stock control—another issue that was crucial for Ford. Concerns about levels of materials and finished goods held in the works were frequently aired at committee meetings, and the desire for efficiency in the organization of the factory resulted in frequent moves toward rationalization and centralization of stock.[47] In February 1919, it was decided to unify factory stores rather than have them distributed throughout the factory.[48] In December, Sir Frank Bowden called for the employment of "a capable Head Storekeeper" to bring about "improvements in storekeeping methods."[49] In April 1920, the company introduced weekly rather than annual stocktaking in order to be able to detect stock fluctuations and to help disclose the company's financial situation.[50] This proposal led to a visit to the Raleigh factory by a Mr. Mellors from Ford's Manchester plant, who advised Harold Bowden that the Ford stocktaking system might not be suitable for Raleigh.[51] The problem of stock organization clearly remained an issue, since as late as April 1928 the committee was considering adopting the system for stocking "sundries" that was in use at Triumph's factory.[52] While no clear resolution of this issue can be deduced from the minutes, its presence there indicates again that Raleigh was engaging at this time with some of the central themes of modernity: rationalization, bureaucratization, and the mitigation of uncertainty (Bauman 1991; Law 1994).

The importance of the Ford Motor Company in promoting these themes is underpinned by its self-promotional activities. Hounshell (1984: 260) points out that from the earliest days of mass production Ford "was completely open about its organizational structure, its sales and its production methods. . . . Ford engineers had no skeleton closets in their factory. Proud of their work, they were anxious to have technical journalists tour the shops and write extensive articles about Ford methods." The effect of this policy was a rapid diffusion of Fordist production technology throughout American manufacturing, and it is consequently no surprise to find that Ford of Manchester offered advice to Raleigh. According to Gregory Bowden (1975: 63), Sir Harold visited Ford's Detroit factory around 1921 to learn more about Ford's production methods. This may be related to reports in October 1921 that Ford was interested in purchasing Raleigh.[53]

Referring to Ford's assembly line and conveyor systems, Bowden (1975: 63) writes that in the early 1920s his grandfather "had not believed the expenditure required to apply these techniques to the Raleigh works to be justified" at that time. Once such equipment had been installed at Raleigh a decade later, though, visits to other companies continued, which confirms my earlier suggestion that Raleigh did not necessarily lead the way among British cycle firms in adopting Fordist production techniques. In February 1933 Raven and his foremen visited the Dunlop works and noticed the cleanliness of the factory, the high pressure of the work, and the excellence of the conveying system.[54] In January 1934, Raven visited the works of the Coventry Chain Company, "where some of the most up to date plant was in operation."[55] Most impressive for Raven seems to have been a 1935 visit to Germany, where he saw "very up to date" equipment at several factories. Most striking was the Union Works at Froendenburg, where "I saw, without doubt, the best equipped and most highly organized factory for repetition work I have ever seen."[56]

Raven's interest in up-to-date equipment meant that during the 1920s, without yet installing moving assembly lines or overhead conveyors, Raleigh did adopt many other Fordist technologies. There are several examples of increasing vertical integration and the purchasing of new automatic special-purpose tools. New production processes were developed to increase productivity, and jobs were subdivided in order to reduce the required skills. Attempts were made also to introduce "scientific" methods to improve the efficiency of production processes and of factory organization. Techniques adopted included research into costing methods[57] and employing an analytical chemist to test materials in production.[58] Raleigh also engaged a number of apparently spurious "experts" who the company hoped could solve certain problems. In May 1928, a man called Hobbs, from a firm called Industrial Psychology, was set to work investigating various aspects of the factory—the internal transport system, "mess room habits," the Sturmey-Archer assembly plant, conditions in the tool room, and lavatory accommodations.[59] Nothing of value appears to have come of Hobbs's work, though. Barely two months after he began, the committee was wondering where he was, and in July it was observed that he was "not doing much useful work."[60] No mention is made of ever receiving his final report. Also during this period, there was a half-hearted and rather comical attempt to introduce strategies reminiscent of the way Ford regulated and controlled its workforce, though without the latter's brutal effectiveness (Beynon 1975; Counter Information Services 1978). In July 1926 the company began to act against thefts of materials from the stores. First a former policeman was hired as a detective.[61] Then, in February 1927, a handwriting expert was consulted to provide psychological information about suspects, "particularly of those about whom we had some doubt"[62]—in other words, it was hoped that he would bolster the committee's own prejudices. When the first results provided no firm evidence against the storekeeper concerned, the committee was "disappointed."[63] The next results were more satisfactory, as the expert's results were "rather derogatory to Worth's character,"[64] but nothing more came of this excursion.

In comparison with the slackness and naivete of Raleigh's attempts to apply Taylorism and Ford-style "sociology," the modernization of the factory was far more serious. This was primarily the work of Raven, as can be seen from the reduced level of expansion during his absence between 1922 and 1929 (although this period might to some degree also mark the boundary between postwar reconstruction and the later modernization program). Much of the new plant bought during the 1920s aimed to make Raleigh less reliant on outside suppliers. This move toward vertical integration was crucial if Raleigh was to achieve anything remotely resembling American practices. Vertical integration was in fact a central feature of the development of the large mass producers in the British cycle industry (Hudson 1960: 88). This tendency can be seen at Raleigh as far back as the period immediately after World War I, when a shortage of supplies of nuts, screws, and similar parts held up production; the solution was to begin producing these parts in house.[65] The company continued to bring parts production in house during the 1920s,[66] and in December 1928 it was announced to the committee that Raleigh no longer had to buy any tools.[67]

Raven continued to push Raleigh toward vertical integration in the 1930s, with the installation of new plant to make pedals, freewheels, and so on.[68] By 1937, the company was claiming that "with the exception of tires and saddles there is practically no component bought outside" (Raleigh Cycle Co. Ltd. 1937: x), and it has become a part of the company's folklore that by the time of the merger with the British Cycle Corporation (which brought Brooks Saddles under Raleigh's jurisdiction) it was producing all the parts of its bikes except the tires and the ball bearings. The vertical integration of the 1920s, though, was far less complete than that of the 1950s. Work was sent outside when to do so seemed strategically wise and cheaper—for example, in 1925 it was done to bypass a dispute with glazers and buffers.[69] This was not, then, a wholesale adoption of "high Rouge" Fordism; it was a more flexible British variant of mass production.

Nevertheless, by the early 1930s, Raven had set Raleigh on a major course of modernization involving a series of "sanctions" for new plant.

Numbered successively in the committee minutes, these sanctions together constituted a continuing process of automation and updating of equipment for tube making, forging, pressing, enameling, and plating.[70] When Sir Harold Bowden returned Raleigh to public ownership in 1934, the share offer claimed that between August 1928 and February 1934 a total of £227,991 had been spent on new plant and equipment (*Times*, February 15, 1934). This bears little comparison, of course, to the $18 million (at least) that it cost the Ford Motor Company to retool for the ill-planned 1927 changeover from the Model T to the Model A (Hounshell 1984: 288, 379n85). It was, nevertheless, a considerable sum, and that share offer was a success. (On February 17, 1934, the *Times* reported that it was oversubscribed.)

For Gregory Bowden (1975: 63), the major feature of Raleigh's modernization program was the introduction of overhead conveyors—the defining characteristic of Fordism in the public mind—in 1931–32. However, the adoption of assembly-line and conveyor equipment was more gradual than Bowden claims. Some kind of mechanical conveying of parts had already been included in provisions for the 1922 factory extension. Just before he left the company, Raven was asked by the committee to obtain quotations for the "pulleys, belting, &c. required to complete shafting arrangements" in the new factory.[71] This may have developed out of Sir Harold Bowden's visit to Ford's factory the previous year, but it may well have been unrelated to Fordism. By the summer of 1928, though, the implementation of conveyors was progressing purposefully. Quotations were obtained and an order placed for a cycle assembly conveyor, which was in operation by December.[72] By October 1930, new layouts had been installed in the frame department and in the bracketing shop.[73] In the next two years, Raven installed a runway in the cycle assembly area, a conveyor in the frame shop, and conveyors linking various workshops and departments to one another.[74] This is clearly the period to which Gregory Bowden refers. However, Raven was still proposing "the possibility of his bringing forward a scheme for a conveyor to carry work all round the factory"[75] as late as September 1935, Until that time, then, conveyors must have been limited mainly to systems within departments, with only a few areas linked together. Nevertheless, the company was claiming by 1937 that its main service conveyors totaled no less than 8 miles in length. In addition to a general service conveyor that automatically delivered parts to their correct stations, there were several localized conveyors in specific areas and in the final assembly area (figure 3.5).[76]

Figure 3.5
A conveyor on the final inspection line at Raleigh, 1935. Original legend: "FINAL VIEWING. Completion of the continual inspection and checking to which every Raleigh cycle is subject in its progress from raw material to finished machine. The viewers are skilled men of long experience, who can detect instantly any variation in adjustment which may need correction." Source: Nottinghamshire Archives.

Modernity and Ambivalence

Even with the above-mentioned conveyors, Raleigh can by no means be regarded as having fully adopted a Fordist approach to mass production. In the minutes from January 1937 there is a telling comment from A. E. Simpson, who became a director of the company the following year when Sir Harold Bowden retired. Simpson is reported to have "hoped the day was not far distant when we should supply to the Depots whatever stocks suited the factory to produce, and the Depots would be automatically stocked by the factory working to a program suited to its requirements, where they could concentrate on one particular type of machine at a time, and whereby the very best use could be made of the conveyors." Most significantly, the minutes continue: "At the present time it is scarcely possible to see two consecutive machines of the same type passing along the conveyor."[77] This comment indicates that Raleigh's adoption of conveyors fell far short of the purposes to which they were put under Fordism. Despite the increasingly high levels of output at this time, rather than engaging in Fordist mass production—"the modern method by which great quantities of a single standardized commodity are manufactured" (Ford 1926: 821)—Raleigh appears to have used conveyors to speed up its traditional production methods.

Might it be fair to say, then, that by not adopting some aspects of American industry in the 1920s and the 1930s Raleigh and other British engineering companies "failed" to adopt Fordism, as many commentators (e.g. Lewchuk 1987; Jessop 1991b; Cooper 1993) argue or at least imply? Were Bowden and Raven mistakenly setting out on an endeavor that they (perhaps unknowingly) had little chance of fully realizing?

It is more likely that their location within the industrial culture of their time and place meant that both men were well aware of the implications of the different paths open to them. The Raleigh committee minutes show that Raven had broad knowledge of manufacturing processes past and present in the cycle industry and elsewhere, and of the alternative methods being developed in other factories and other industries. Bowden was a major industrialist with seats on several national committees. He was also fully aware of the direction in which manufacturing in general was moving. The industrial strife of the 1920s meant that both men were regularly reminded of the state of industrial relations both in their own industry and in those with which they had to deal commercially, while Raven's close contact with the shop floor and Bowden's welfarist paternalism allowed the Raleigh committee at least to guess how various options might affect the relations of production in the company.

It seems fair to assume, then, that Raleigh's adoption of mass-production did not involve a mistaken understanding of what elements constituted "Fordism" or American mass production. It arose, rather, out of a rational assessment of what elements of that approach could be usefully and successfully adopted and incorporated within a British context. Bowden and Raven would have been all too aware that the British market for bicycles was not comparable to the American market for cars and thus could not sustain the kinds of radical restructuring of shop-floor organization and industrial relations that took place at Ford. They would also have known that British labor would not accept the conditions of employment imposed at Ford, nor would they willingly exchange their control of the shop floor for day rates; in any case, Raleigh could not afford day rates high enough to sway them in this.

Anyway, the "flawed Fordism" thesis assumes that the Raleigh management might have believed that their own strategies of the previous few decades had been inferior to those of American manufacturers, despite the fact that these strategies had made Raleigh a highly successful and still-expanding company.

The changes that took place at Raleigh and elsewhere in the British cycle industry during the 1920s and the 1930s were, then, elements in a strategic and conscious transformation of the sociotechnical frame of the bicycle that borrowed from other industries and other frames. This borrowing took place within the context of an understanding of the ways in which these new elements were likely to interact with those aspects of the previous frame that remained in place, because the major players in this transformation knew they would not be able—and perhaps did not even want—to dislodge them. They also knew how compatible these new elements were likely to be with the existing cultures of cycle production and cycle use. Thus, what seems from one perspective to have been a "flawed" version of Fordism should be more fairly regarded as a successful (in the short term, at least) transformation of production methods that avoided jeopardizing the industry's relatively good relations of production—a considerable risk had they adopted "true" Fordism. This underlines the specificity that must be applied in setting out the features of any sociotechnical frame, since the way technologies, practices, social groups, and events interact may overlap from one frame to another but will not be identical. Which features will carry across from one frame to another cannot be assumed a priori, and what may be deemed appropriate or desirable in one setting might be highly inappropriate elsewhere.

This interpretation of the modernization at Raleigh brings to the fore something that is often understated, if present at all, in accounts of the industrial changes taking place in Britain at this time: the ambivalence about modernity and modernization that was present alongside the apparently partial adoption of Fordist practices. The general ambivalence toward Fordist mass production within British industry during the interwar years is epitomized in the writings of Sir Harold Bowden, as Cooper highlights. Cooper quotes an article from July 1922 written by Bowden after his visit to Henry Ford the previous year:

> The traditions of Great Britain have at once been a help and a handicap. They have sometimes prevented us from taking full advantage of new ideas. They have been to some extent responsible for the reputation we hold of being slow to move. But a craftsman does not readily discard a tried and proven tool for anything that is new. He fears for his reputation. Therefore he will often let the younger men

experiment . . . the British working man is getting over the effects of the strain of war and will face any competition that may come. But he will face it by skill and craftsmanship. He will take the best machinery, the best ideas, the best material from any quarter, and from the whole produce something which only the craftsman can ensure.[78]

This comment reiterates the contrasts I have identified between Bowden's approach and Ford's (for example, with regard to the scrapping of old equipment), and this was far from the only time that Bowden expressed such a view in public. In another article from 1922 (or possibly the same one from a different syndicated source), he wrote even more damningly of mass production:

Let me first clear away one misconception of many writers on this subject. There is a wretched combination of words thrown at our heads whenever this topic is mentioned. We are told that "mass production" is the solution of our troubles. In our case the ideal would be to put some metal in at one end of a machine and turn out a bicycle at the other. In my opinion, "mass production," as a mere general phrase, is quite unsuited to British methods. As a manufacturing nation we are proud of our craftsmanship. . . . We can produce on a big scale, but let it be with the craftsman's work.[79]

Nine years later, Bowden was no longer praising traditional methods above the modern. In "A message to industry" (1931) he wrote that it was "not enough merely to keep the wheels of industry set in motion, if we are to re-establish ourselves in the prosperity for which we are hoping." He continued: "More than ever it is essential to bring to current problems a mind attuned to the demands of the present day and not allow progress to be hampered by bygone prejudices and methods. For the manufacturer, there is the vital need to employ a wider perspective than ever before and to cultivate a deeper appreciating of the requirements of his customers, whether at home or abroad." At the same time, though, he persisted in extolling the virtues of craftsmanship. He continued by calling "for the workman to do his part by reviving the spirit of craftsmanship in his work which, in the past, has made our products second to none in the world's markets."[80]

Cooper (1993: 14–15) points out that by the 1920s "there was no place for craft production techniques at the Raleigh factory." He regards Bowden's notion of the craftsman as "an almost mythical character, evoked [in his 1922 rejection of mass production] for emotive effect in his justification of Raleigh's enforced retreat from the demands of international corporate capitalism." He links the inability of Raleigh (and other British engineering firms) to fully introduce Fordism with a general retreat from industrialization and urbanism that typified "the sentiment of influential elite culture." He quotes a Raleigh advertisement from 1923 that contrasts the urban world of work with a description of country lanes, "bluebell-covered banks crowned by hedges" and "the spire of the village church," arguing that this "cultural iconography . . . because of the introspectiveness it engendered, so damagingly undermined the foresight of early twentieth century British industrialists."

Cooper's distinction between material and cultural production, in which the latter (in the form of an advertising text) is seen as a way of masking the failings of the former, overlooks a genuine ambivalence toward modernity that has characterized cycling ever since its earliest days (Patton 1993). The negative side of this ambivalence is perhaps best expressed for the earlier part of the twentieth century in the illustrations of Frank Patterson, whose vision of Britain has been described as "one of quiet village streets, uncluttered by cars and commercial signs, of half-timbered cottages and tranquil churchyards," in which "touring cyclists, wearing plus-four breeches, fall asleep under trees by the peaceful roadsides and their machines are generally single speed, often have a full chainguard, and never seem to be locked up" (*Bike Culture Quarterly* 2, 1994, March: 44). (See, for example, figure 3.6.) Behind these idyllic scenes, though, modernity was crucial both as the stimulus that led city dwellers into the countryside to begin with and as the force that provided them with affordable and well-built bicycles and with metaled roads on which to ride them. Clearly this paradox was already a central problem for Raleigh back in the interwar period, and in ways more complex than Cooper's portrayal of Sir Harold Bowden cynically manipulating images to hide his company's deficiencies suggests. In the early 1930s, with the installation of conveyors progressing, Raleigh was proud to be able to reduce its prices "not by cheapening the product, but by utilizing the most scientific methods of production, [by which] the Raleigh quality is altogether maintained."[81] The 1934 Raleigh share offer stated that "the plant, machinery, tools and equipment are modern and efficient, a large proportion thereof having been installed during the past few years," and that "up-to-date methods of mass production, including modern conveyors, automatic machinery, and machine tools, are employed" *(Times*, February 15, 1934). Sir Harold Bowden had, by this time, evidently begun to find his love of efficiency compatible with the new ideas he had condemned a decade earlier. Nevertheless, the company was forced to address the discontinuity between its traditional approach and the modern techniques it was beginning to embrace. While no longer using craft

Figure 3.6
Frank Patterson's drawing of the George and Dragon, a pub near Horsham. Originally published in *CTC Gazette*, December 1947. Reproduced with permission from *Britain's Counties by Frank Patterson: Sussex* (GM Designs, 1992).

methods of production, Raleigh was still offering non-standard designs in the 1920s—a feature of the British cycle industry that had already seriously hindered moves toward the American system decades earlier (Harrison 1977; Lewchuk 1987). By the 1930s, as Hudson points out, this practice was all but abandoned by the large producers, leaving custom provision to the smaller firms (Hudson 1960). Thus in 1933 Bowden told Raleigh agents that the new assembly system made it difficult to continue producing non-standard designs. This was to be remedied by the introduction of a wider range of models.[82] In 1928 there had been ten different models, five of which were available also with a woman's or a girl's frame. By 1937, Raleigh was producing fourteen different models; 20 years later, at the peak of its output, it was producing seventeen or eighteen models.[83] This expansion of the range to make up for the shortcomings (from Raleigh's point of view) of mass production explains in part the unusually large range of models still produced by Raleigh into the twenty-first century. Similar trends are suggested by the 1930s catalogs of Raleigh's major competitors of the time, BSA and Hercules.[84] Paradoxically, while these larger companies expanded their product ranges to compensate for the standardization caused by the introduction of new technology, smaller manufacturers using traditional production methods were able to offer model ranges just as wide. For example, a Dawes catalog of the late 1930s lists 21 models—more than BSA, Hercules, or Raleigh offered.

Expanding and Constraining the Sociotechnical Frame of the Mass Bicycle

It is clear from the above discussion that the transformation of the sociotechnical frame of the bicycle in the 1920s and the 1930s had profound implications for the nature of the artifacts that the larger manufacturers were able to produce in the following decades. While the European craft system had lent itself well to customized individual machines, automated technology increasingly pushed production toward larger numbers of nearly identical products, albeit distinguished by styling. Fordist equipment, in however un-Fordist a manner it was used, eliminated customizing from the repertoire of the large manufacturers—which by the end of the 1950s, after two decades of absorption of smaller companies by Raleigh and the British Cycle Corporation, meant approximately 80 percent of British production.

The new sociotechnical frame constructed in the interwar years provided great—if short-lived—prosperity for Raleigh and its few mass-producing competitors. Raleigh's production more than doubled between 1922 and 1930, and by 1940 it had increased again by a factor of nearly 3.5. This expansion was curtailed only by World War II, during which cycle production was cut to 5 percent of factory output as most plant was taken up with making munitions (Bowden 1975: 74). During the postwar years, though, the company rapidly built production up again, reaching a million cycles per year in 1954 and an all-time pre-merger peak of 1.3 million in 1956.

Raleigh's share of British cycle production rose from less than 10 percent before World War I to about 17 percent during the 1920s and the 1930s. In the postwar period, and as the new technology was consolidated and expanded during the 1950s, Raleigh's production grew to about one-third of the UK's output, bolstered in part by absorbing so many of its competitors. Just before the 1960 merger with the British Cycle

Table 3.3
Raleigh's production relative to total number of bicycles produced in UK. Based on table 3.1 above and on figures from the Bicycle Association of Great Britain (1991: 7).

	UK total	Raleigh
1907	624,000	28,156 (4.5%)
1912	528,000	46,945 (8.9%)
1925	640,000	98,612 (15.4%)
1930	660,000	110,074 (16.7%)
1935	1,997,000	355,210 (17.8%)
1940	1,000,000	356,387 (35.6%)
1945	656,000	179,144 (27.3%)
1950	3,528,000	820,388 (23.3%)
1955	3,562,000	1,047,846 (29.4%)
1960	2,278,000	1,197,585 (52.57%)

Corporation, Raleigh had more than 50 percent of UK production. That rose to more than 75 percent after the merger (*Financial Times*, April 19, 1960). (See table 3.3.)

Modernization and expansion of productive capacity have been central to the progression of sociotechnical frames in the cycle industry. Clearly these features are not limited to economic concerns about outputs and projected incomes, and clearly they are crucially influenced by relationships formed within and beyond companies. Thus, in addition to internal tensions between the advocates of change and of financial prudence, there is a high degree of "social learning" about change that takes place across a range of related firms (Rip et al. 1995), shaped by discourses and practices that are both industry-specific and more widely dispersed within business culture. It is through this social learning that sociotechnical change becomes a process of the joint development of shared—or cooperatively learned—visions, and not just an outcome of commercial imperatives.

4

Working for Raleigh: Sociotechnical Change and the Relations of Production

A wealth of literature exists in the sociology of work and industry that has been barely addressed within social studies of technology. Much of this work emerged out of the debates on the labor process sparked by the publication of Harry Braverman's *Labour and Monopoly Capital* in 1974, and it is of relevance to SST in its exploration of the relationship between technological and industrial change. Writers have sought to examine the impacts on labor and on the workplace of the technological and broader changes that began to take place across a wide range of industries during the 1960s and the 1970s (see, e.g., Zimbalist 1979; Wood 1982). What is especially valuable in many such works is that the often explicitly Marxist approaches of their authors, while sometimes bringing problems of their own in narrowing the scope for analysis, frequently lead them to see technological change as something situated within broader social and political processes—an approach from which SST would do well to learn.

These works consequently provide a valuable point of comparison to set against the few studies that have been made of work and technology in SST (e.g. Noble 1984; Anderson 1988; Saetnan 1991; Mort and Michael 1998; Mort 2002). Such comparisons are necessary in part because both stages in the "product life cycle" that critics of social studies of technology pinpoint as its main, overly narrow, focus—i.e., design and use (Orlikowski 1992; McLaughlin et al. 1999)—are located at some remove from the kinds of issues addressed by labor process literature. This chapter is an attempt to follow writers such as Noble in highlighting the importance of the relations of production as a component of sociotechnical change. In the case of Raleigh, the relationship between technology and labor was crucial in the construction and subsequent destabilization of the sociotechnical frame of the mass bicycle.

Management and Workers under the Bowdens: The "Raleigh Family"

Raleigh has generally been portrayed as having good relations between management and workers, especially up until the 1960s merger. Bowden (1975: 202–211) describes the company as an enlightened employer, with the entire workforce, plus dealers and agents throughout the world, encompassed in the company's eyes within one big "Raleigh family." Raleigh was, indeed, one of a number of benevolent employers within British industry (see Smith et al. 1990; Bradford 1996) before its merger with TI, providing a welfare officer and surgery for employees from the 1920s and a convalescence home from 1936 (Bowden 1975: 203–205). The company bought staff a sports ground in 1930 (ibid.: 205), while the new offices built on Lenton Boulevard in the same year included a concert hall, a dance floor, and a "cinematograph theatre" for the use of all employees (*CTC Gazette* 40, 1930, no. 5: 151). These enlightened welfare policies are closely related to Sir Harold Bowden's paternalistic attitude toward his staff—an approach also taken by other companies of the era, such as Cadbury's (Smith et al. 1990) and Royal Enfield (Bradford 1996). Bowden's position on labor relations meant that he made a point, as had his father, of getting to know employees personally through regular walks around the works (Bowden 1975: 211) and of making himself available to anybody with a grievance. This accounts for his first-hand knowledge, displayed occasionally in the committee minutes, about the morale of the shop floor.

Evidence of a cooperative relationship between management and workers can be found in interviews from an oral-history project carried out in Nottinghamshire in 1982–1984. There are positive—if sometimes also rather baffling—comments on relations between workers and management, and between workers and foremen, from a retired trade union official, a retired Works Director, and a retired foreman, all of whom worked at Raleigh from the 1930s on.[1] The retired foreman describes Sir Harold Bowden as "a benevolent sort of bloke. Okay he'd got to run a firm and he'd got to make a profit, but he did take an interest in the employees."[2] The retired Works Director was William Raven's replacement, Leonard Clarkson. In his interview, he sees the interest taken by management in their employees' social life as the source of what he perceives as Raleigh's "good record" in the eyes of its employees.[3] Even the trade unionist Jack Hallam, who had been branch secretary for the National Society for Metal Mechanics between 1955 and 1981, feels that the foreman in the wheel-building shop he worked in after the war "wasn't unreasonable." Unlike his disciplinarian "Victorian" predecessors, this was one of "the second generation" of foremen, who eventually even came to accept the existence of trade unions.[4]

Workers' accounts of Raleigh often identify the shift in 1960 from a family business to a corporate one as signaling a change in industrial relations and in the atmosphere of the factory. A 1964 article about the labor disputes of the 1960s attests to the generosity of the Bowdens, in contrast to the post-1960 management.[5] The retired foreman mentioned above remembers the Sturmey-Archer gear shop in the 1930s as having a "marvelous atmosphere . . . we had more fun there than going to the Empire [Music Hall]." He reminisces: "But oh life was great. And we hadn't got a lot of money and often we were laid off. But all in all we all made fun of each other and everybody took it in the right spirit."[6] This rose-tinted perspective is echoed in the positive memories of the 1950s recalled by a trade union convenor I interviewed in the early 1990s. This man, who had started work at Raleigh in 1949, remembered the 1950s as happier than the present, with a sense of community in the factory that had now been lost:

> When we talk about the old days, and this is a fact, this is, when there was that camaraderie between everybody, and Friday was a fun day, Friday was. People used to knock off work early on a Friday and just sit down talking football, then what they're gonna do at weekend, everything like that. And there were a lot of singing going off, go round in any department, and loads of groups of people would be singing, and they'd almost see it as Friday, the weekend, get out the place, sort o' thing. And you don't get that now. Everything seems so serious now, believe it or not. They've got the serious part o' life. I don't know whether it's all the debt people's into . . . it's that much more serious nowadays.

Others interviewed for the oral-history project remember things differently, though, indicating that an individual worker's perspective was shaped in part by the particular shop they worked in. One man describes conditions in the factory during the 1920s as "a bit hard really," though "not too bad" in the motorcycle frame shop where he worked himself. He describes the conditions in the pitch house, though, as "horrible," continuing: "Well you can imagine it was like. . . . The nearest to the Black hole of Calcutta I've ever imagined you know. But er . . . it were terrible. I only went in there as a visitor. I didn't work in there. The people who did work in there their skin turned yellow, all over their face, hands, They were in a shocking conditions they were."[7] Working at Raleigh during this period is frequently described as dangerous and unpleasant. Another man describes being in the turnery during the 1940s as a "soul destroying job," telling how "your hands were sore, and, you'd no protection in those days, you know, you used to wrap your hands up in Sellotape, well, not Sellotape, the old black tape."[8]

Most striking, perhaps, is an anecdote told by a man who worked in the press shop as a boy in the 1920s. He first describes how he was taught to use the equipment: "They put me on the press, showed me what to do, and then when I'd finished that job said 'Well now I'll show yer how to alter it,' and you had to do all your own tool setting, no guards, no guards on the machines." He goes on to tell of an accident he was involved in, emphasizing the conditions in the shop: "It was dangerous, well I'll tell yer and er I was doing this job one Monday and one of me mates cum and he wanted. . . . In yer back stave where you put yer back wheel in, I was slotting them and you had what you call a liner in there stiffner, well he wanted one of these liners flattening and slotting wi'out being in this thing and he held it wi' his fingers while I done it, and I chopped his fingers straight off."[9] While this man recognized it had been partly his friend's fault for not using the stave, when he was called to the office the following day to explain what had happened he put the blame on the lack of guards on the presses. This or a similar incident was in fact noted by the management committee in May 1925. The minutes tell of a "very unfortunate accident [that] occurred . . . when an operator in the Press Shop had the thumb and first finger of his left hand cut off by the press which he was working. The press was properly guarded by a device supplied with the machine."[10] The company's firm response here, and the lack of any further investigation, reflect the realities of life for those members of the "Raleigh family" who were dependent for their welfare on management benevolence: few health and safety regulations, and little trade union authority over staff conditions.

Piece Rates and Resistance on the Shop Floor

A story that can be traced through several years of management committee minutes links the issues of shop-floor authority, technological change, and rates of pay. Something that comes across in several of the oral history interviews is Raleigh's long-standing reputation in Nottingham as a very poorly paid factory,[11] a situation which seems to have been common in the cycle trade in the Midlands (Bradford 1996) and is still felt to be the case by workers today. A number of the individuals interviewed for the oral-history project talked about the piece-rate system that was in operation until the late 1960s, particularly a man who went to work at Raleigh in the late 1930s. He describes his job in the old turnery making parts for pedals, for which he earned a penny per hundred: ". . . I'll just tell you what I did. You stood with a pair of wooden pliers made, put your thing in pushed it on to a drill, a reamering drill both ends, dropped in, picked another one up, put it in the [?] pushed it on, now that was the job and you stood there at the machine doing that all day long, you just had your break for dinner. There was no such thing as a 10 minute tea break, you had tea but you had it in the old fashioned mash can as we called it."[12] This man's father was a chargehand at Raleigh, and he describes the way the piece-rate system worked: ". . . no one really had a set wage at Raleigh. The only set rate he got was a ceiling rate if you follow what I mean because if he earnt over that, although he earnt over it, Raleigh never paid it him."[13]

As chargehand, his father was responsible for distributing wages:

> Whenever you did any work there you made a note of it, of how many you'd done of each particular work and the code number, and the price of how much a thousand or how much a hundred, whatever the case may be. At the weekend you gave your chargehand your slip of paper with the amount of work you'd done that week on the Friday. Now my father then would come home on the Sunday with all the employees underneath him, and I'd say he had roughly 60, 70, or 80 people under him maybe more, counting the underchargehands men, he would then have his Sunday dinner and when we was at Sunday School he would go into the front room, get his ready reckoner and paper and would work out each amount of work for each individual, how much they'd earnt for that particular week.[14]

This responsibility allowed the man's father to manipulate the system: when there were workers with earnings above their ceiling that couldn't be paid to them, his father would use this to supplement the wages of others who had earned little that week, rather than give the money back to the company.[15]

The problem of exceeding the ceiling is referred to by another man who worked in the motorcycle gear shop as a youth in the 1920s. He tells how he "became so proficient I began to make more than my quota and was warned by my workmates to slow down because I might be clocked and the agreed payment reduced."[16] The same man describes apprenticeship at Raleigh as consisting of "proving your ability to operate the various machinery and jobs at a fraction of the wages till the age of 21 or 22, when, if not then sacked [you would be] given the full engineers' rate."[17] The son of the chargehand also makes the point that once a boy got to be 16 he'd often be sacked and replaced by a 14-year-old who could be paid less: "This was how it used to go on all through Raleigh during the thirties. . . . They were after cheap labor with the children."[18] (See figure 4.1.)

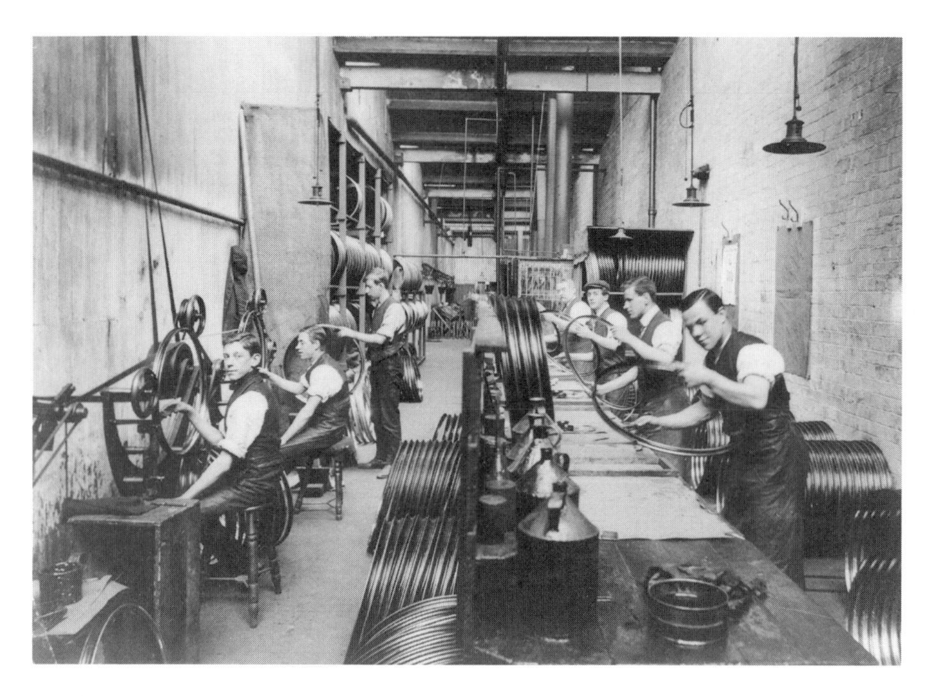

Figure 4.1
A pre-1940 photo of boys and men in the rim-lining department. Source: Nottinghamshire Archives.

The issue of piece rates is highlighted in Alan Sillitoe's 1958 novel *Saturday Night and Sunday Morning*, which focuses on a 1950s turnery worker in the Raleigh factory, where Sillitoe had himself worked during the 1940s (Bowden 1975: 10). Ceilings on earnings don't seem to figure here (either because of artistic license or because they had ended by Sillitoe's time). The novel's protagonist, Arthur Seaton, takes advantage of the piece-rate system by earning as much as he can get away with. Seaton is careful to stick to the set production rate of 100 parts per hour when the rate checker is watching (Sillitoe 1958: 27). For most of the day, though, he works at his own pace, producing 1,400 parts per day— ". . . if you went all out for a thousand in the morning you could dawdle through the afternoon and lark about with the women and talk to your mates now and then" (ibid.: 26).

Seaton regards his working conditions as a great improvement on those before the war: "Now, and about time too, you got fair wages if you worked your backbone to a string of conkers on piecework." (ibid.: 23)

Nevertheless, the danger of having his rate cut is ever present—a point he is warned of by the foreman (ibid.: 52–53):

> "How much this week?"—though like all pieceworkers he knew the exact number of pound notes folded into his packet.
>
> Robboe spoke in a hushed voice: "Fourteen. It's more than the tool-setters get. I'll be in trouble one of these days for letting you earn so much. They'll be lowering your price if you're not careful."
>
> It was a subtle warning, and Arthur braced himself against it, saying gruffly: "Not if I know it, they wain't."

Piece rates have been discussed in several ethnographies of the shop floor. Lupton (1963) writes of "the fiddle" practiced by piece-rate workers in the 1950s who would claim wages for work they hadn't actually done, often with implicit connivance at lower levels of management. This issue has been taken up by both Burawoy (1979, 1985) and Harris (1987). Harris (ibid.: 66–71) argues against Burawoy's position that the fiddle constitutes "consent" among workers to the existing relations of production and the authority of management. Instead, Harris suggests a more complex interaction of shop-floor practices in constructing the relations of production between different groups of workers as well as between workers and management. She explores the conflicting world views that exist not just between supervisors and "ordinary workers" but also between workers in two different plants of the same chemical firm. This supplements Burawoy's concern to demonstrate the multiple roles and relationships of certain categories of workers, such as foremen and works managers, who mediate between management and labor. Law (1994) makes a similar point in his account of the hierarchical ordering he found among different categories of workers in a scientific research laboratory.

The minutes of the Raleigh management committee meetings support this line of argument, notably in the positioning of William Raven between the committee and the shop floor. This is brought out also in the case of Tivey, a worker at Raleigh who makes a great many appearances in the minutes through the 1920s and whose story shows how integral the relations of production are to an analysis of the *sociotechnical ensemble* of the bicycle. This story provides an account of one man's manipulation of the piece-rate system. The committee's methods of dealing with him indicate a subtle and gradual change that can be detected in Raleigh's approach of the 1920s, reflecting the shift that was taking place from a sociotechnical frame whose production methods were based on the

European factory system to one based on the mass production of standardized products.

The Case of Tivey: Resistance and Innovation in the Relations of Production

Tivey's story can be told only partially and somewhat speculatively. Tivey appears to have been a 1920s real-life equivalent of *Saturday Night and Sunday Morning*'s Arthur Seaton, but in tracing his story in the minutes it becomes clear that they provide a biased and partial account of his life and of his role in Raleigh's history. Tivey is referred to more than forty times between October 1921 and September 1930, yet at no point is his first name given, and nowhere is it made clear what his exact position is in the company. Since he was a member of the works staff rather than of management, I have also come across no other reference to him in my research on Raleigh outside the committee minutes. This highlights the problem of telling the histories of those who aren't in privileged positions in society (Thompson 1980; Anderson and Zinsser 1988), and it also harks back to my difficulty in tracing external references even to William Raven, a long-standing member of the committee who later became a director of the company. Tivey's place in the committee minutes is nevertheless an important one, serving as a reminder of workers' resistance to the use of scientific management and new technology as a means of redirecting the labor process. This resistance, however difficult to reconstruct, helps to counter the technologically determinist assumption of writers such as Braverman (1974) that workers subjected to changes such as the introduction of scientific management accept them unproblematically. It also gives an early indication of one of the factors that would lead later to the destabilization of the mass bicycle sociotechnical frame.

The basis of Tivey's resistance was money, in particular the amount of money the committee was happy for him to earn through piecework. The first reference to Tivey, in October 1921, sets the tone for the entire decade of his appearances in the minutes, dealing with "a conversation which [Raven] had had with Tivey in regard to a suggestion that the latter should go on piecework." "It was felt," the minutes record, "that such an arrangement would result in considerable saving and Mr. Raven was therefore requested to endeavor to arrange satisfactory terms."[19] A month later it was reported that Tivey was holding out for a piece rate of one shilling per wheel, which the committee agreed to in order to bring his shop under piece rates "at the earliest possible moment."[20] The wider

Figure 4.2
A pre-1940 photo of foreman and workers in the wheel shop. Source: Nottinghamshire Archives.

context of this encounter was the slackening cycle trade of 1921 (discussed in chapter 3), which led to a serious cut in production and jobs at Raleigh. That Tivey was not sacked at this point, and that he got his way when a 20 percent cut in the piece rate was being implemented across the factory,[21] including foremen's wages, suggests he was of some value to the committee.

Tivey is mentioned in the minutes only a few times over the next 5 or 6 years, but these references begin to give some clues as to his status. In March 1922 it is reported that he was "guilty of sweating the girls under his control." Consequently, "it was decided that the company should undertake direct payment to each operative and that Tivey should be put back to ordinary fixed wage." This decision seems not to have been acted upon. Rather, Tivey is said to have given his "assurance that he is now paying his people on a higher scale."[22] In July of the following year the minutes mention that Tivey had agreed to a reduction on his piece rates of 1d per pair of cycle wheels and 2d per pair of motorcycle wheels.[23] In March 1924 it is reported that his wages over the previous eight weeks had averaged £12 14s per week. As a consequence, Tivey was to be "called upon to accept a reduction of 1d per pair of wheels but such reduction was to be borne exclusively by Tivey and his two sons and was not to be passed on in any way to the other work people in his Department."[24] In November 1925, Tivey's rates were cut from 10d to 8d per pair of wheels, giving an expected 50s saving per week. From this meeting on it was also decided to note Tivey's wages in the fortnightly minutes.[25] Three months later, Tivey's earnings for the previous fortnight are quoted as £19 per week, with the comment that "it would appear from the list of wages supplied that the boys in the Wheel Dept. are getting much too little."[26] Two weeks on, his wages were down to £10 10s, and arrangements were being made with Tivey for a man to sweep the shop floor "at Tivey's expense."[27]

It begins to become apparent that Tivey was a foreman, or possibly a chargehand, in the wheel department, and that the committee believed he was exploiting the works staff under him. Although the company was clearly unwilling or unable to fire him, the committee was keen to find ways of bypassing the apparent abuse of power. The method it used to do this was direct intervention between foreman and works staff; this illustrates the paternalism of Raleigh under its European style of management as well as the strong position of foremen under this system (Littler 1982).

Tivey's later appearances in the minutes show a subtle change of tactics on the part of the company in response to what appears to be his growing defiance. By the early part of 1927, Tivey's average weekly wage dropped to about £9, hitting a low of £6 4d in April.[28] By the following meeting it had risen again to £11 6s, prompting Holland (the Works Manager at that point) to seek a further reduction in his piece rate. Piece rates were again being reduced throughout the factory at this time, which may explain why the committee was unhappy with Tivey's ability to earn high wages.[29]

Most significantly, at the end of May the committee sent Tivey and a man named Ridgeway to Coventry "to inspect a special wheel-building machine introduced by McKenzie."[30] While it was more than a year before any new equipment of this sort was up and running in the Raleigh works, this decision marks a change of approach toward Tivey that belongs less with the European factory system than with Fordism—in particular, with what Hounshell (1984: 252) identifies as the Fordist concern "to eliminate labor by machinery" (see Ford 1926). It also recalls Marx's observations about the use of machinery to eliminate labor resistance (1970: 436): "It would be possible to write quite a history of the inventions, made since 1830, for the sole purpose of supplying capital with weapons against the revolts of the working-class. . . ." At the same time, the way this new labor-saving equipment was introduced exemplifies the ambivalence and awkwardness of Raleigh's move toward mass production.

The installation of new wheel-building equipment was delayed because other purchases were given priority.[31] In the meantime, Tivey's rates were again cut, but he continued to defy the committee by maintaining high earnings. In late November 1927, he earned an average of £11 2s per week. There is a dry comment in the minutes that "apparently he has not yet felt the recent reduction on motorcycle wheels."[32] In January 1928 his earnings reached £14 10s 3d, which led to a further reduction of ½d per pair of wheels. Consequently his wages dropped to £9 13s, but they had gradually climbed up again to £14 3s 6d by the end of March.[33]

During this same period, the new wheel-building machine was being assembled, although from the committee's perspective this was happening with little cooperation from Tivey. In June it was reported that "Mr. Holland has come to the conclusion that Tivey has been willfully retarding the progress in connection with the new wheel-building machine and consequently has handed over that work to Moseley, the foreman of the Tyre Stores."[34] A month later it was handed back to Tivey although Holland thought he wasn't "doing the job justice." Consequently, Tivey was to be given "a further trial" with the machine.[35] By the end of August he was earning a rate of 6¼d per pair of wheels, and the committee aimed to reduce this to 4d per pair.[36] By October, Tivey and the new wheel-building equipment were no longer being listed under the "Works Staff" heading in the minutes; they had their own separate heading, under which it was reported that Tivey had earned £600 in 1927 and £500 the previous year.[37]

The wheel-building equipment installed in 1928 appears to have been of Raleigh's own design, since in 1929 a new machine was fully installed by McKenzie, the man Tivey had visited in Coventry. The McKenzie machine arrived in January 1929, but by February the committee was already dissatisfied with it. McKenzie had promised to oversee installation and sort out necessary alterations and repairs, but this was delayed for several weeks due to illness, and it wasn't until July that the machine was finally declared satisfactory. At the same time, Raleigh was talking to another manufacturer in Germany, and by December it was in the process of buying a machine from Birmingham.[38]

Tivey continued to obstruct Raleigh's attempts to force his wages down and perhaps even displace him with new technology. His wages fluctuated a great deal during 1929, ranging from £5 7s 6d to £12 2s and averaging out at just under £9 per week. That Tivey continued to be uncooperative is clear from references made during the spring of 1929 to some rims that were damaged in the drilling operation; the cost of repairing these by welding was deducted from Tivey's wages in mid April. (He nevertheless managed to earn £11 4s during those weeks.) In July he was forced to accept a rate of 4d per pair of wheels.[39] In March 1930 he was finally put on a fixed wage of £7 per week. That September, Raven was still "not at all satisfied that Tivey was doing his best for the company."[40] It isn't clear from the minutes what became of Tivey after this final reference. Nevertheless, the Tivey case is extremely valuable in supplementing the anecdotal evidence from the Nottingham interviews and the factory episodes in *Saturday Night and Sunday Morning*, both in giving an idea of what it was like to work for Raleigh and in offering an understanding of the company's relations of production.

Paternalism and Control in the Ordering of the Mass Bicycle

The case of Tivey supports Harris's (1987) argument that our understanding of the relations of production should not be restricted to relations between workers and management as two unified blocks (see also Feldberg and Glenn 1983). Rather, the relations of production at Raleigh involved, in addition to the divide between workers and management, an overlapping structural differentiation among different groups of workers and among different sections of management. As the earlier discussion of Bowden and Raven illustrates, the views of the former would always override those of the latter, highlighting Raven's position somewhere between management and workers. The Tivey case shows, moreover, that during the 1920s management was willing and able to intervene directly between a (presumed) foreman and the workers under him, illustrating the reduction during this period of the absolute authority over the shop floor enjoyed earlier by foremen (Littler 1982).

The positions of workers also were differentiated among the different categories of men, women and girls, and boys. Hudson (1960: 127) points out that the seasonality of the cycle industry and "the light repetitive work that was a consequence of widespread mass production" meant that the industry had had a greater proportion of female workers than elsewhere in light engineering as far back as the 1890s. This supports Glucksmann's (1986) argument about the underacknowledged role of female labor in the development of mass production, although there is little reference to women workers in the Raleigh minutes apart from an occasional comment on their wage levels.

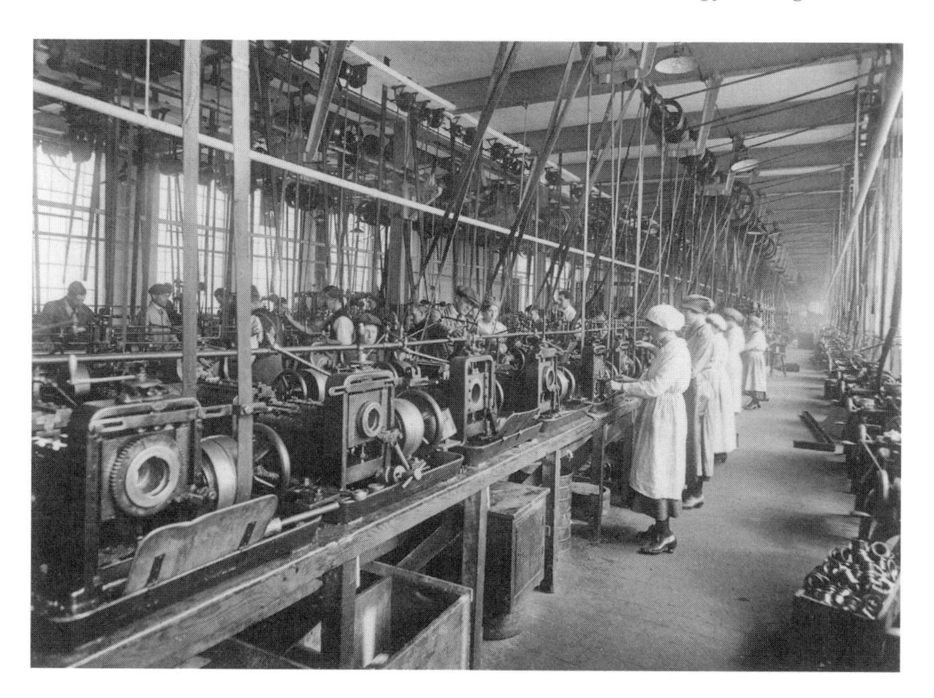

Figure 4.3
Women and men in the factory, 1921. Original legend: "Cycle Road Factory, Lower Room, May 13, 1921." Source: Nottinghamshire Archives.

The relations of production at Raleigh before 1960 comprised, then, a complex mixture of often contradictory practices and discourses, paralleling Bowden's simultaneous attraction to both tradition and modernity. Alongside the welfare paternalism that gave the Bowdens a reputation for benevolence could be found the more conventional capitalist values that demanded cheap labor and a compliant workforce as well as the efficient use of labor-saving equipment.

Nevertheless, workers themselves as well as management identify paternalism as the ,most significant aspect of Raleigh's attitude toward its employees. This draws on the notion, already set out, that Raleigh was a "family business," a perspective backed up by comments from several interviewees in the Nottingham oral-history project. The committee meeting minutes highlight this by showing that it was often Sir Harold Bowden himself, rather than a delegate, who was asked to deal with difficult workers, whether in the case of Tivey damaging cycle rims or in that of an employee named Dury who had been seen "talking to some women during business hours."[41]

Despite its paternalism, though, the Raleigh committee was clear in its views about the rights and mutual expectations of workers and management. Without more detail about the Dury case, it may not be fair to judge the committee for what seems a rather extreme response. Other examples in the minutes show, though, that it frequently made such attempts to keep workers in their place. At one meeting a complaint was made about the workforce's being served tea at 9 A.M. and 4 P.M. on the ground that "opportunities for refreshment were gradually being turned into additional mealtimes." This example shows the extremes of pettiness to which the committee (more likely Bowden himself) was willing to go to save money: "It was decided that a particular case of misuse be found, warning given, and if after that the misuse should continue the privilege would be withdrawn."[42]

In industrial disputes, too, whatever Bowden said publicly about his policies, Raleigh was in no hurry to give way to trade unions' demands. This was particularly the case during the period immediately after World War I, a period of general labor unrest throughout the country and not just in the cycle trade (Hudson 1960). Raleigh's introduction of piece rates across the factory at this time met with some resistance among the more skilled elements of the workforce. Consequently, there was surprise in the committee when its decisions were occasionally met with cooperation—for example, when repair hands backed down on their demand for skilled rates and accepted piece rates in May 1920.[43]

Perhaps the reason for Raleigh's relatively peaceful industrial relations lay in a reluctance at this stage—among the committee and among many trade unionists—to engage in direct confrontation (something that would plague the company under different circumstances during the 1960s and the 1970s). Back in the 1920s, the committee responded to a dispute with glazers and buffers that began in September 1925 first by asserting that Raleigh paid higher rates than its competitors did and then by sending work outside. Eventually, in April 1926, it came to an agreement with the new Brassworkers Union shop steward that there would be no further drop in wages provided glazers increased their output.[44]

Raleigh's ambivalence toward the more entrenched positions that characterized other sections of British industry at this point is illustrated by the distance it maintained from management in disputes elsewhere. In June 1919, with the first wage cuts and the introduction of piece rates, it was felt "inadvisable" for Raleigh to join the Engineering Employers' Federation (EEF), although a month later the company committed itself to making a verbal (but not financial) expression of support for other employers in the event of strikes.[45] When such strikes occurred, though, especially in coal and steel production, the committee was far more concerned with the effects on its own business than with the plight of its fellow employers. In 1922, members of the EEF initiated a national lockout in an attempt to assert management's right to "direct control" over issues such as overtime, staffing, bonuses and apprenticeships (Friedman 1977: 200). For Lewchuk (1987: 110), this was a crucial element in the EEF's strategy to achieve control of the level of workers' effort and consequently of production, a strategy that "doomed British engineering" to what appears to him an inevitable collapse in the 1970s. However, it was business as usual for the members of the Raleigh committee, who were not members of the EEF—although the lockout is mentioned briefly in the minutes.[46] The same is true of the General Strike of 1926, which affected Raleigh only on the first day, bringing out 37 percent of the workforce. The committee was less concerned with these workers' striking, though, than with the effects on the factory of a shortage of coal and transportation caused by strikes elsewhere.[47]

One way of understanding the apparently contradictory discourses operating at Raleigh at this time is provided by Law's (1994) notion of "modes of ordering." Law distinguishes four such modes in operation at the Daresbury Laboratory, where he carried out his research: *enterprise, administration, vision,* and *vocation.* These modes of ordering, which characterize the approaches of the different groups of people Law encountered, occasionally "butt up against one another" in specific places in the laboratory. There might thus be a degree of conflict between a dynamic manager whose approach is structured by enterprise and vision and more administratively focused actors (ibid.: 123). The Raleigh minutes show that vision, vocation, and enterprise were important modes of ordering there too.[48] Sir Harold Bowden and William Raven embodied different constellations of these three modes—Bowden's vision, for example, lay in his leadership role in the company, while Raven's lay in pushing the company toward modernization.

Beyond such modes of ordering, the sociotechnical frame of the mass bicycle was based on four interlinked pairs of opposing values in constant tension:

- the production values of craft and mass production
- the consumption values of leisure-related and transportation-related cycling

- the discourses of "tradition" and "modernity," and associated discourses of anti-urbanism and the city
- the industrial relations of paternalism and managerial control.

These four sets of oppositions together shaped the construction of the new sociotechnical frame of the bicycle in the mid twentieth century, and the constant tensions generated by this may explain the underlying instability that Lewchuk attributes to British versions of mass production.

Framing the Transition: From the Factory Bicycle to the Mass Bicycle

The Raleigh committee minutes are an excellent case study in the transformation that took place generally within the cycle industry after the 1890s boom and especially in the 1920s and the 1930s. They help pinpoint this period as a transitional phase between two sociotechnical frames of the bicycle. The earlier of these two frames was centered on elite culture and on the approaches of the industries from which the nascent cycle manufacturers had emerged (e.g., carriage making and the production of sewing machines and small arms). The later frame, centered on mass (though not quite standardized) production and consumption, was consolidated during the 1940s and the 1950s on the basis of the modernization that had been pushed forward by actors such as William Raven with the cautious backing of their bosses. This transitional period saw a change in priorities within Raleigh. Concerns about the quality of individual products were overridden by the technical constraints imposed by new production equipment. Pride in the skills of company craftsmen was replaced by pride in the factory and its output. And the company's role within the industry came increasingly to be defined in terms of market competition and expansion rather than the qualities of the firm or its products.

At the same time, there were also many continuities. The company's increasingly aggressive market strategy continued to be rooted in the same financial outlook that had informed the price consciousness of both Bowdens throughout Raleigh's history. The management control that was facilitated by new organizational and production methods had also been a constant goal of their hands-on approach, though operationalized in a very different manner than under strict Fordism. Most significantly, the running of the company during both periods was based on a set of relationships and skills that remained consistent from one sociotechnical frame to the next—from Frank Bowden and G. P. Mills to

Harold Bowden and William Raven, the company's progress was largely directed by a partnership of entrepreneurial and engineering skills. This partnership, with all its accompanying tensions, is crucial to an understanding of the changes taking place at this time, not least because of the way its operation within the sociotechnical frames of both the factory and mass bicycles was embedded within the twin roles of Managing Director and Works Manager.

The events that took place during this first transitional period offer a valuable insight into sociotechnical change "in the making" (Latour 1987). The protracted debates in the Raleigh minutes between cost cutting and modernization, the changing model ranges and prices of both Raleigh and its competitors, the simultaneous collaboration and competition that led to the industry's adoption of new methods, and the question of how to intervene in the relations of production show, together with Sir Harold Bowden's social, political, and economic commentaries, how important features of the earlier sociotechnical frame began to be superseded. By the late 1930s, the sociotechnical frame of the factory bicycle, centered on small-scale manufacture using skilled craftsmen, was no longer adequate to describe the situation in the industry. The experience of small craft-based producers was no longer the main focus of the industry as a whole. Rather, their experience had come to be defined by their relationship to larger producers, and such relationships among different social groups were just one aspect of how the new frame was considerably more complex than the earlier one.

Some features of the new sociotechnical frame derived directly from the preceding frame—a situation that explains the apparently paradoxical continuity inherent in sociotechnical change (Bijker 1995). Change itself lay in the juxtaposition of these established elements with new ones derived from other frames—notably the incorporation of new production technology and organizational methods from American industry. This is in fact one of the two mechanisms Bijker identifies that can bring about sociotechnical change—the clash between two existing frames that sparks the establishment of a new one. Bijker's other mechanism is also evident in this transitional period, in that entrepreneurs and engineers such as the Bowdens and William Raven did not have a background in the pre-1870s engineering trades or even in the establishing period of the early cycle industry. Both Raven and Bowden came into the industry when the transition from one frame to another had already begun (although only just so), and neither had been socialized into the earlier frame. This raises interesting questions about the periodization of sociotechnical frames and about inclusion and exclusion within them. The transitional period between the sociotechnical frames of the factory bicycle and the mass bicycle lasted 20 or 30 years, and Raven and Bowden both spent considerable parts of their careers in the industry during this phase (as did many others). The significant changes that were taking place throughout the industry were facilitated, then, by a general absence of people in the cycle trade who had any great degree of *inclusion* in the older sociotechnical frame. To an extent, this situation mimicked the openness of the 1870s in allowing a great burst of innovation—this time not in product design, but in production methods, production processes, and industrial organization. This openness and innovativeness—in whichever form—has remained a constant feature through all the sociotechnical frames of the bicycle. Figure 4.4 depicts the playing out of Bijker's two mechanisms of sociotechnical change during the transition from the factory to the mass bicycle—a transition brought about through an encounter among innovators and entrepreneurs influenced by American approaches, new groups of users on an increasingly larger scale, and the contradictory discourses of modernity and anti-urbanism.

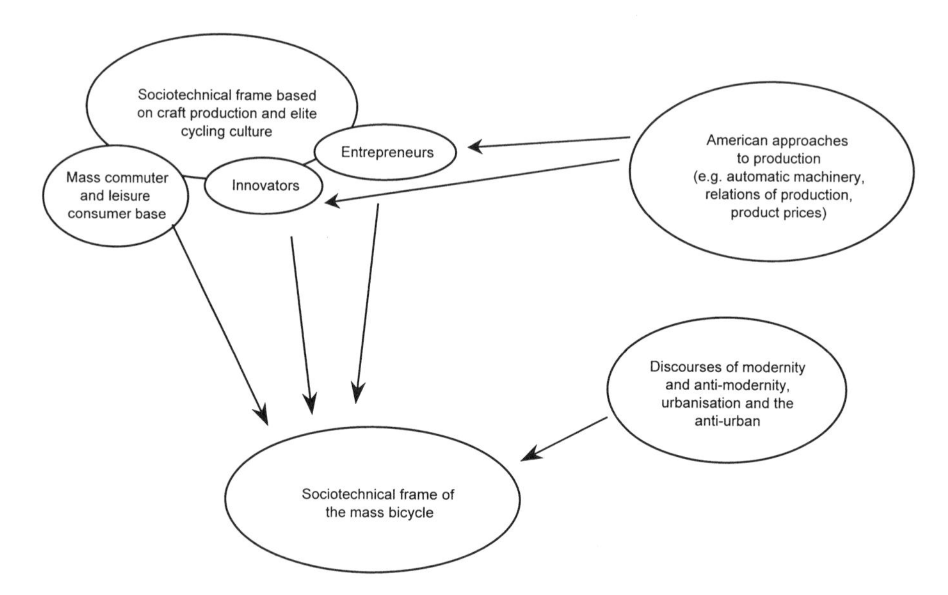

Figure 4.4
Encounters leading to the construction of the sociotechnical frame of the mass bicycle.

These new consumers and discourses reflected a change within the *culture* of cycling—a change characterized by increasing ambivalence about modernity.[49] However, ambivalence about modernity was by no means peculiar to Raleigh or even to cycle *manufacturers*—indeed, as I will show in chapter 6, it continues to be a defining feature of cycling culture. Although I have not discussed it in any depth regarding this period, the uses to which bicycles have been put have played an important role in shaping (and have themselves been shaped by) the different sociotechnical frames of the bicycle. Up until the boom of the 1890s, cycling was almost exclusively a leisure pursuit of the wealthy. It became more popular largely because competition among producers led to the reduced prices of the mid-1890s bicycle boom, and this facilitated a shift toward the use of bicycles as a means of mass transportation. Consequently, the expansion of production in the early twentieth century both fed and was sustained by a growth of utility cycling (McGurn 1987). The presence of a strong utility-based market as an element of the sociotechnical frame of the mass bicycle was another feature that changed significantly as that frame too began to be superseded. The changing culture of cycling was an important part of the destabilization of mass production.

5

Destabilizing the Mass Bicycle: Technological, Organizational, and Industrial Change

In chapters 3 and 4, I examined the process of modernization during the interwar years that established the sociotechnical frame of the mass bicycle. By the time World War II began, this process was largely complete, so Raleigh and other manufacturers were able to make a significant contribution to the war effort and then, after the war, to move quickly toward the production peaks of the 1950s. The end of this golden age of the cycle industry was already in sight, though, even while production and sales were unprecedentedly high. The two dominant producers—Raleigh and the British Cycle Corporation—found their export markets saturated during this period of high prosperity, which effectively put a stop to any further global expansion. This happened at the same time that the increasing incomes of consumers at home led to a rapid rise in car ownership and a shifting cultural perception of bicycles no longer as transportation but as toys or, at best, leisure equipment. The period from the late 1950s to the mid 1980s can be seen, consequently, as a second, and far more complex, transitional phase for the bicycle, and especially the British bicycle industry. First, there was a slow and painful restructuring of the industry's plant, capacity, and product ranges, and also of its relationship to retailers and consumers. Second, consumers themselves were undergoing significant changes in terms of demography, wealth, and cultural perceptions of the bicycle and of other modes of transportation. Accompanying these changes were struggles within the bicycle industry between management and labor that mirrored similar struggles elsewhere in British industry at this time (Lewchuk 1987). Finally, there were changes in the global economy, both within and beyond the cycle industry, that seriously hindered its ability to solve by itself the internal problems that were affecting it.

There were several points during this transition that could easily have led to an irreversible closure of most British cycle production. However, the tenacity of the industry's participants combined with two crucial developments to push a declining industry toward a new sociotechnical frame, which would prove able to survive more effectively in the economic, technological and cultural climate of the 1980s and the 1990s. The two developments were (1) a set of radically different methods of manufacturing bicycles that complemented new organizational practices borrowed from Japanese and American companies and (2) the rise of a new style of bicycle, the mountain bike, that dovetailed neatly with these new methods and embodied new cultural trends that helped revitalize cycling as an activity. These factors also facilitated the transformation of the bicycle—at least, the bicycle as sold in the developed world—into a product based on global networks of manufacturers and suppliers.

Raleigh's Merger with the British Cycle Corporation

The 1960 merger of Raleigh with the British Cycle Corporation (BCC) appears at first to have been an odd move, in view of Raleigh's success at the time. In the 1950s, Raleigh's share of the market was generally larger than the BCC's except in South America (Bowden 1975: 98). However, the saturation by the two companies of the entire world bicycle market, combined with a loss of home sales as private automobiles came within the scope of working-class wages, brought about a crisis in the industry that was deeply felt by both companies. From the mid to the late 1950s, Raleigh's production remained fairly static at approximately 1.1 million bicycles per year. While the factory expansion of 1952 had enabled a slight increase in capacity, the opening of No. 3 Factory in 1957 proved to have been ill-conceived. This was the first time in the company's history that such an ambitious move failed to pay off. The new factory remained empty for several years (ibid.: 94) as the output of the industry as a whole dropped from 3.5 million machines in 1955 to 2 million in 1958.[1] In light of these factors, it is no surprise that the two conglomerates began to realize that the world cycle market was finite, and that if they weren't to destroy each other through competition they would have to form trading agreements.

As early as 1953, talks on the possibility of sharing the South African market were held.[2] In 1959, Raleigh agreed to close down its South African factory in return for an interest in that of TI (Bowden 1975: 98). This agreement came about quite amicably, according to Gregory Bowden, although Raleigh's committee minutes and the BCC minutes of the mid to the late 1950s indicate a slightly more tense relationship. In

1955, two different proposals for cooperation between the two companies were put forward by the BCC, but both failed to proceed. The first of these was that Raleigh should curtail production of bicycles while the BCC should do the same for variable gears.[3] This would have left the BCC as the dominant British cycle manufacturer and Raleigh as the dominant manufacturer of three-speed hub gears; had the agreement gone ahead Raleigh would almost certainly have disappeared with the rising popularity of derailleur gearing in Britain during the 1960s (Hadland 1987). These discussions broke down, but they were followed by another proposal—also never fully pursued—that the BCC would give up manufacturing variable gears if Raleigh gave up manufacturing steel tubing, and that each should buy its entire requirements for these items from the other.[4] While receiving provisional support from the Raleigh committee, this proposal was eventually rendered irrelevant by the 1960 merger.

It appears from the above examples that it was generally the BCC that initiated discussions with Raleigh, perhaps out of a need to impress the TI Board. From the minutes of the TI "Cycle Division Council" (i.e., the BCC management) it is clear that the latter was answerable directly to TI for much of its decision making. The Council Chairman is continually minuted as needing to refer to TI's Chairman in regard to particular actions—for example, whether they should buy up an additional cycle company.[5] The willingness to merge with Raleigh appears likewise to have been underlain by anxieties about the Cycle Division's relationship with the TI Board. For example, in April 1959, a year before the merger with Raleigh, the BCC Council was concerned about the need to be seen to be "again operating profitably" in time for the next TI Board meeting.[6]

The apparent uncertainty at the BCC benefited Raleigh quite considerably. The Board of the new post-merger Cycle Division was dominated by former Directors of the pre-merger Raleigh,[7] of whom George Wilson—the final Raleigh Chairman—was made both Chairman of the new Cycle Division and a Director of TI. Furthermore, it was the Raleigh name rather than that of a BCC brand that was retained for the group's overall identity, although other brand names continued to be used. Most significantly, as the new Cycle Division began to consolidate in the early 1960s it was Raleigh's Nottingham site that became established as its headquarters, and eventually as the center of all production. Nevertheless, the 27 years that Raleigh spent as TI's Cycle Division proved the most troublesome period of its history, marking an all-time low in labor relations and the company's first serious financial losses since before World War I. During the 1970s, especially, the company began to lose sight of its reputation for product innovation, while only at the end of that decade did it begin to take the kinds of innovative initiatives it had previously been known for in terms of production.

The events leading up to the merger reinforce the sense of homogenization among manufacturers that was a feature of the sociotechnical frame of the mass bicycle. If Raleigh's goal was to be the largest cycle company in the world, as it sometimes claimed to be in its catalogs before 1960 and as it certainly was afterward, it was not alone. During the 1920s and the 1930s, as I showed in chapter 3, Hercules, BSA and other companies were matching or exceeding Raleigh's rate of modernization. Hercules, Raleigh's main competitor, concentrated all its efforts on expanding its output. Hercules produced its 3 millionth machine, "built complete" in under 10 minutes, in November 1933 (*Times*, November 22, 1933), whereas Raleigh's 3 millionth machine didn't come off the conveyors for another two or three years,[8] despite the fact that Raleigh was the older company. After its sale to TI in 1946, Hercules was associated with an expansion of brands that was just as ambitious as Raleigh's in the same period. The BCC deal brought under Raleigh's wing an array of brands that had been acquired by TI during the previous few decades, including Sun, Norman, Phillips, and Hercules. Similarly, when Raleigh bought BSA's cycling interests in 1957, these also included the New Hudson and Sunbeam brands, bought by BSA in 1944 (Hudson 1960: 175).

The 1960 merger was, then, a merger of two strong industry actors sharing similar commercial goals, but sharing also an awareness that their international markets were no longer expanding. The merger constituted an alliance between two almost equally placed firms, both in a precarious position that made them mutually dependent and mutually cautious. This relationship can be understood from a number of different viewpoints, with no clear-cut winners or losers in terms of economics, company autonomy or brand identity. The negotiations certainly favored Raleigh over the BCC in many ways, since the latter was making losses during the late 1950s. TI therefore needed cooperation with Raleigh to secure the future profitability of its Cycle Division. Ironically, it also needed to ensure Raleigh's survival in order to protect the market for its steel tubing subsidiary, TI Reynolds (Bull 1991: 31)—hence the proposal to give up producing variable gears if Raleigh would give up producing tubes. After the merger, the BCC immediately lost its identity. In the following years its factories were sold and many of its brands discontinued.

Things were not all positive for Raleigh, though. It had spent £5 million on an unused factory and hence had "a desperate shortage of cash"

(Mansell 1973: 83). It therefore needed the resources of the £260 million TI group as much as TI needed Raleigh's brand names and markets. After the merger, Raleigh retained its identity but lost its autonomy. This had serious consequences, since Raleigh's subordination to the TI Board in decision making almost led to the Cycle Division's closure at a number of points during the 1970s and the 1980s. From TI's point of view, the merger had been hoped to stave off the potential losses that might have resulted from further competition of its Cycle Division with Raleigh, and it might even have hoped to profit from Raleigh's reputation. In fact, it turned out to have acquired a company that it would prove unable to develop satisfactorily, hampered by production facilities that were fast becoming outdated and volatile labor relations that would soon come to a head once declining sales brought to light the company's underlying structural and organizational problems.

The merger constituted a striking conclusion to the process of homogenization that had been in progress within the industry for several decades. It was, though, a far less clear-cut affair than Gregory Bowden (1975) implies when he treats it as a case simply of Raleigh successfully incorporating yet another of its competitors. Rather than this being, in actor-network terms, a case of one actor enrolling another and making itself integral to the other's activities (Callon 1986b), which is close to what Bowden depicts, there was instead a mutual interdependence between Raleigh and the BCC—something which could be more usefully explained in the framework of "social worlds," where different groups collectively construct a project through cooperation on shared objectives (Clarke and Montini 1993; Fujimura 1992). The maintenance of this relationship, through the agreements of the late 1950s and the subsequent merger, was felt on both sides to be essential if the whole industry were not to collapse.

Bounding the Frame: Innovation and Organizational Change under TI

The way Raleigh sought to re-establish its market position in response to the various hurdles it faced in the 1960s and the 1970s is rarely explored in social studies of technology. Rather than grab wildly at every opportunity that presented itself, Raleigh pursued a careful restructuring of relationships within the industry, seeking in fact to exclude a significant number of former allies. Mort and Michael (1998) write of the simultaneous *disenrollment* at Vickers's Barrow shipyard of products, other companies, and workers during the 1980s as a part of the company's efforts to focus on its "core business" of defense contract work, and this example has some relevance to the TI Raleigh situation. Using actor-network terms, Mort and Michael describe the "proactive process of excision" (ibid.: 358) from the company's network of (1) a non-military naval technology with considerable market potential, along with the civil engineering company that had jointly developed it, and (2), some years later, of workers who had been taken on to produce the Trident nuclear submarine but who were no longer needed as that contract drew to a close. Mort and Michael describe these disenrolled actors as *phantom intermediaries*; they are absent physically from the network, but they retain a ghost-like presence that both signifies paths that might have been taken and serves to discipline those who remain (ibid.: 392).

The business strategies of TI Raleigh in the 1960s and the 1970s included similar excision policies to those of Vickers, intended to re-establish the company's identity among consumers and hence reverse the decline caused by lost export markets. Attempts to delineate more clearly the brands that constituted the Raleigh product range and the dealers that made up the "Raleigh family" were accompanied by the removal of models, brands and dealers that were felt to hinder the company's revival. This process was underpinned by a great deal of effort (and money) spent on developing innovations, initially at the level of new products, but also in terms of the organizational structure of the company, marketing strategies and the production process itself. This broad range of innovations was developed alongside, and partly as a result of, increasingly tense industrial relations that culminated in a six-week strike in 1977. The loss two years later of major overseas markets such as Iran and Nigeria made the late 1970s a defining period in Raleigh's decline within the TI group.

New Products and New Markets

The product innovations of the 1960s and the early 1970s followed from what was perhaps Raleigh's most foolhardy mistake: its rejection, in 1959, of Alex Moulton's design for a small-wheel bicycle with a suspension system. After Moulton's unexpected success as an independent producer, Raleigh launched its own equivalent, the RSW16, in 1965.[9] In a 1973 interview, Raleigh's chief designer, Alan Oakley, distinguishes Raleigh's small-wheel models from Moulton's in terms of their "essential" focus, asserting that Raleigh focused on the small wheels whereas Moulton focused on the suspension (Mansell 1973: 88). This isn't the way Moulton himself recounts his design process. In a 1979 lecture, he recalls the

Figure 5.1
A girl on a Raleigh Chopper, from Raleigh's 1971 "Happy Families" brochure. Source: Nottinghamshire Archives.

choice of small wheels as coming first, since "in all other vehicles the reduction in size of wheels has been in the direction of design evolution because so many advantages generally speaking flow from it" (quoted in Roy 1983: 68). Suspension for Moulton was simply a means of countering the extra shock experienced with smaller wheels, drawing on his experience designing the suspension in the Mini car.

Despite the great success of Moulton's design, it was Raleigh that developed the concept of small-wheel bicycles across a broad market range, revitalizing the ailing cycle industry. The RSW16 was followed by the RSW14, the Raleigh 20 shopping bike, and (in 1970) the Chopper, designed in response to a craze for dirt-track roadster bikes that emerged in the United States in the late 1960s and opened up a new market in children's bikes. In 1967, as a result of Moulton's financial and production problems, Raleigh bought Moulton out and began to produce the Raleigh Moulton.

This new range of bicycles, which all featured the same basis of small wheels on a unisex frame, marked a shift in Raleigh's philosophy toward a new "underlying realization that bicycles are consumer goods, not bits of light engineering" (Peter Seales, head of marketing at that point, quoted in Mansell 1973: 85). This philosophy has been roundly condemned by cycling advocates who see Raleigh's approach to marketing since the 1960s as endorsing the idea that bicycles are not a valid form of personal transportation. Andrew Ritchie (1975: 174) laments that "[cycle] manufacturers seem to have capitulated to the car along with everybody else," with the Chopper as "the worst aspect of this collapse of confidence in the real value of bicycles." The Chopper, he asserts, "is not really a bicycle anymore"; it is "a luxury consumer toy" (ibid.: 177). Similar sentiments were expressed in more depth by Kath Hamer at York Cycleworks, a cooperatively run cycle shop with a strong commitment to promoting cycling. In the following quotation from my interview with her, she is talking not about the Chopper but about a bicycle seen by many as its 1990s equivalent: the Raleigh Activator:

> I think it's the most horrible thing on earth, it represents everything that is bad about the way Raleigh sells bikes. It's unpleasant to ride, very poor quality because they've tried to build something to a price that you can't make for that price. It weighs a ton. I think it would put kids off bicycles. It's a bit like the small-wheeler; they produced that; I had one when I was younger. You have an RSW 16 with the real fat wheels—basically, Moulton invents a bike that's got suspension on, skinny wheels, and they're real titchy and it rides really well. Raleigh then crush him, buy the name, produce a bastardized version of it with huge wheels, and the whole thing feels like riding through treacle. And then you produce what is in effect with the Activator a kid's toy, not a bike. The kids go—oh, it's got that suspension that's just brill. The thing that they're aspiring to, the full-flown superlightweight mountain bike with Manitou-2s [a brand of suspension forks] and everything, that's the image they've got. They're then given this thing that weighs 45 pounds—nearly 40 pounds, that was a bit exaggerated. But it is very heavy. The forks will need to—I doubt the forks will last long. . . .
>
> Basically the thing doesn't measure up to the dream at all. And I think it then puts kids off. Kids'll get that when they're about 12-ish. They'll ride it for two or three years. As soon as they're 15 they'll be aspiring to motorbikes, and they'll just associate bikes with that great big donkey thing they had with suspension on. It's not something that's a pleasure to ride. And you can get kid's bikes that are. And it's this concept that Raleigh is selling toys rather than bikes. And I think it's very short term from their point of view. And it's one reason why they can't sell nice bikes, because people associate them with rubbishy bikes. . . .

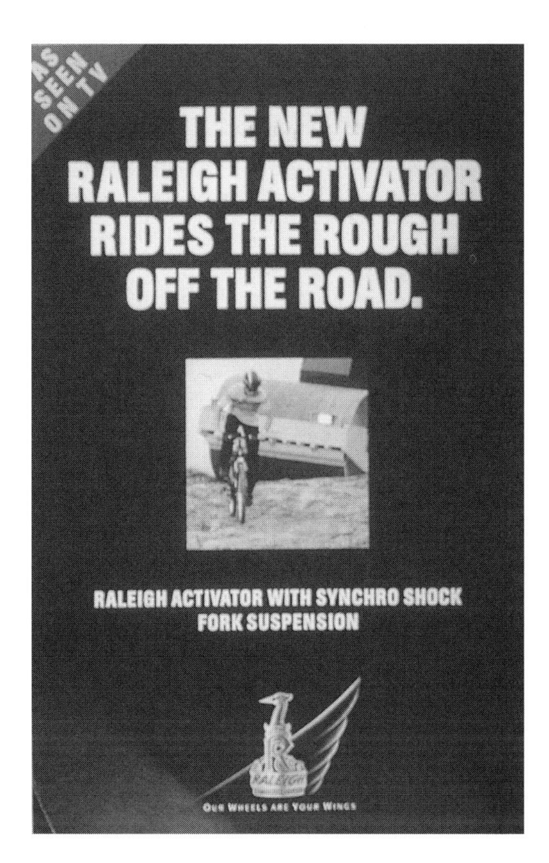

Figure 5.2
A promotional poster for the Raleigh Activator.

> I think it's a shame. I think it just puts people off cycling, it just panders to everything that is a pity in the market.

The Activator, a front-suspension mountain bike, was launched in 1992 with the deliberate aim of bringing suspension down to a more affordable price—it sold for about £200, whereas the cheapest suspension bikes sold for about £450. It was followed in 1993 by the full suspension (front and rear) Activator II. Both models were discontinued a few years later once the fad for suspension was seen to have diminished. Underlining the marketing—rather than cycling—rationale behind the Activator, Yvonne Rix, the Marketing Director responsible for it, told me "We're a marketing company." "Our market," she continued, "is still very child- and teenager-oriented." It is no surprise, then, that for Kath Hamer the Activator epitomized the negative aspects of Raleigh's marketing strategy since the

1960s, and as such was a bad advertisement for utility cycling. She felt it had especially negative implications for the chances of establishing a continuity of cycling from childhood through to adulthood. The Activator is seen, then, like the Chopper before it, as reinforcing an attitude that the bicycles is a child's toy rather than a form of transportation. Even Raleigh's own advertising (figure 5.3) suggests, jokingly, that the "natural" progression is from a bicycle to a motorbike and then to a car.

The new "consumer product" philosophy at Raleigh, and the product innovations that philosophy was used to promote, were accompanied in the 1960s by a restructuring of the company in terms of marketing, product management, consumer research, dealer support and training, stock control and sales management (Mansell 1973: 85). These changes were designed to address a problem that had arisen out of the merger with the BCC: Raleigh now owned nearly twenty different brand names,[10] a situation perceived (by TI, at least) as compromising Raleigh's brand identity (Bowden 1975: 99). Consequently, between 1962 and 1972 the number of model specifications available was reduced from 900 to 172, "but with a wider range of types of bicycle" (Mansell 1973: 85). At the same time, the company launched new public relations exercises and advertising campaigns, and restructured its dealer network: "We weeded out most of the cottage industry until we had 1,500 aggressive dealers left." (Peter Seales, quoted in ibid.) The motivation behind these activities was the belief that "the products have to be pulled through the dealer by the appeal to the consumer, and the dealer has in effect to become a loosely affiliated employee of the manufacturer—hence the trend in the car industry to seek single-franchise dealers" (Mansell 1973: 85). In other words, Raleigh's approach to sales and marketing, from the 1960s on, was focused, in actor-network terminology (see Callon 1986a, 1986b), on the goal of establishing Raleigh as an *obligatory passage point* for other actors and *actants* linked to the cycle industry, notably products, retailers, and consumers. By developing new marketing strategies alongside popular innovations, the company attempted to bring about a situation whereby consumer demand would reinforce the perception of dealers that they must join the Raleigh dealer (and actor) network. This strategy, involving the adoption of completely new modes of ordering (Law 1994), entailed first finding new markets to replace those that had been lost overseas. These losses had arisen initially through the saturation of the world market in the 1950s, but also through the rise of domestic manufacturing in Raleigh's traditional overseas markets. In the late 1970s, as imports into the United Kingdom

Figure 5.3
A 1993 advertisement for the Raleigh Pioneer.

were rising, Raleigh also lost its three main export markets through factors external to the cycle industry. In the United States, owing to the increased strength of the pound against the dollar caused by the policies of the new Conservative government, it became uneconomical to import bicycles from Britain. In Iran, the 1979 revolution led to a ban on all Western imports. In Nigeria, Raleigh's biggest African market, the oil crisis of 1979–80 forced the government there to cancel all external payments, including those owed for orders from Raleigh.[11] These events are significant in that they are the same "external shocks" cited in the shift from Fordism to post-Fordism (Harvey 1989).

Raleigh responded to its loss of markets in part by attempting to generate domestic demand for new products through innovations such as the small-wheel bicycle. It also began to redirect its marketing toward Europe rather than the former colonies it had previously depended on. In 1982, Managing Director Roland Jarvis told the *Guardian* : "We have given up our global ambitions."[12] Between 1975 and the mid 1980s, a major element of the company's new strategy was supporting a Tour de France team, which won 77 stages.[13] Raleigh's European sales increased during the late 1970s and the early 1980s, providing a good return on the £250,000-per-year investment.[14] Sponsorship of the team, then, was an indirect strategy used by Raleigh to enroll consumers and dealers.

Testing the Boundaries: Disenrolled Actors and the Monopolies and Mergers Commission

As a part of the project of reversing its fortunes, Raleigh instituted a more structured relationship with its dealer network, bringing dealers more closely into the "Raleigh family" described by Gregory Bowden. Aside from the increased support from Raleigh's sales staff, dealers were now regularly brought to Nottingham to meet the staff, tour the factory, and learn of new developments; "above all they are made to feel at home, to feel that they belong" (Bowden 1975: 208). One particular means used by Raleigh to draw dealers deeper into its network was the establishment in the 1970s of favored dealerships to specialize in the sale and servicing of Raleigh products. "Five-Star Dealerships," accounting for some 14 percent of Raleigh's UK sales in the early 1980s, received—in return for dealing exclusively with Raleigh—preferential discounts and help in upgrading premises, training staff, and advertising (Office of Fair Trading 1981: 13–14). They also benefited from the business of servicing Raleigh bicycles bought through mail order or from dealers unable to provide after-sales support.

Although it was developing mechanisms to consolidate its dealer network and its consumer market, events of this period show Raleigh to have become far removed from its expansionist outlook of earlier decades. At the same time that some dealers were being brought more firmly into the network, others were excluded—"weeded out," in Peter Seale's words. This aspect of Raleigh's strategy is highlighted by investigations of the company made by the Office of Fair Trading and by the Monopolies and Mergers Commission in 1980–81 (OFT 1981; MMC 1981), which resulted from complaints made against Raleigh by "multiple" retailers to whom Raleigh had refused to supply bicycles (MMC 1981: 21, 32).

The investigations into Raleigh were a test case for the Competition Act of 1980 concerning anti-competitive practices, and the results were expected to have repercussions for other industries where manufacturers refused to supply "cut-price" retailers.[15] They were also a test of Raleigh's distribution policy, by which it agreed or refused to supply bicycles to retailers according to the following criteria: the retailer's geographical location relative to existing dealers, whether the retailer was believed to be likely to sell bicycles as a loss leader, whether it could provide technical advice to customers, whether it could provide servicing facilities, whether it would stock spare parts, and whether it could demonstrate a long-term full-year commitment to selling cycles.[16] Raleigh also claimed to take a retailer's creditworthiness into account (OFT 1981: 16), an element of its distribution policy that was not questioned by either report. Nevertheless, the OFT found Raleigh's distribution policy an "anti-competitive practice" (ibid.: 34) and referred the case to the MMC, which investigated the matter further, taking into account two things that the OFT could not: Raleigh's claims that it needed to protect its brand image and that its well-being as a company was in the public interest (MMC 1981).

Raleigh's rationale for refusing to supply certain retailers was based on protecting its existing dealer network. A new dealership would not be opened that might jeopardize trade for an existing dealer in the same location. This policy was intended to retain dealers' loyalty by supporting their ability to make a living and to ensure the longer-term survival of the cycle industry by protecting the existing structure of its retail side. Fears were expressed by Raleigh and by supporters of its distribution policy who gave evidence in the two reports (these included the Bicycle Association, representing manufacturers, the National Association of Cycle and Motor Cycle Traders, and the Cyclists' Touring Club) that to supply multiple retailers would not only provide unfair competition for its existing dealer network but also endanger dealers' livelihoods through price cutting. This was argued to be a threat to safety, since the multiple retailers' ability to undercut specialist cycle dealers relied in part on their lack of pre-sale service checks and after-sale support. In turn, Raleigh expected (and some dealers confirmed) that supplying multiple retailers would result in resentment from dealers who might turn to imported bicycles, thus jeopardizing Britain's cycle industry, of which Raleigh constituted the major part. These last two points formed Raleigh's "public interest" argument (OFT 1981; MMC 1981).

Raleigh's distribution policy was in fact not new. Cycle retailers had responded angrily to Raleigh's decision to start supplying bicycles to the retailers Curry's and Halford's as early as 1936.[17] Such policies are still in effect today, not just at Raleigh but throughout the bicycle industry and other industries. York Cycleworks, for example, was refused supplies of Raleigh bicycles when it opened in the early 1980s, since Raleigh already had a major dealer in the center of York and two "five-star" dealers close by. However, in the early 1990s that shop was the York dealer for eight mountain bike manufacturers. Just as Raleigh had refused to supply the Cycleworks, partly as a result of agitation from its existing dealers, those eight manufacturers of mountain bikes would consult with their main York dealer before opening a new account in the area.

Clearly the outcome of the MMC's investigations had not affected the workings of the cycle industry, even a decade and a half later, despite the fact that, while accepting geographical location as a valid criterion (MMC 1981: 38), the MMC regarded Raleigh's policy toward multiple retailers as anti-competitive (ibid.: 37). Nevertheless, while the company was ordered to begin supplying bicycles as requested, its point about brand image was accepted, and it was not compelled to supply the Raleigh brand. Rather, it could supply any brand of equivalent quality provided it allowed retailers to publicly identify these brands with Raleigh (ibid.: 42–43). This was something Raleigh already did with a number of non-discount mail-order companies (ibid.). In effect, then, Raleigh won the case despite the ruling against it, to the extent that the company's Marketing Director during the 1990s remembered the MMC as having ruled in Raleigh's favor. Even at the time, the Managing Director treated the case not as a defeat but as an opportunity to redress the company's troubled recent history and expand into the new markets offered by multiple retailers.[18]

This case and the revised marketing strategy that formed the background to it demonstrate how far Raleigh's heterogeneous engineering continued to extend beyond the bounds of the company's 64-acre factory in Nottingham even after it had abandoned its "global ambitions."

The Transformation of Raleigh: New Technology and the Reconfiguring of Industrial Relations

Raleigh's industrial-relations problems of the 1960s and the 1970s led to changes in the production process that (although less haphazard) recall the changes of the 1930s. While the evidence for the earlier period suggests a fairly piecemeal approach to expansion and modernization, during the 1970s and the 1980s Raleigh restructured its systems of manufacturing and management in a far more self-conscious way, drawing on many of the ideas of post-Fordism and flexible specialization. This situation lends support to the view of Giddens (1990) and others that "high modernity" is characterized by a greater degree of reflexivity than earlier forms. This section thus offers a point of comparison between the modes of ordering used at Raleigh under mass production and those used in the shift toward something else.

The cycle industry's troubles in the late 1950s had led to redundancies, reduced work hours, and strikes at Raleigh before the merger with the BCC was even announced.[19] Once the new Cycle Division was formed, though, trade unionists had good cause for concern, despite realizing that the merger was "probably a good thing for the cycle industry."[20] As part of its post-merger rationalization, Raleigh began to close its newly acquired factories, in Birmingham in particular, and to move production to its underutilized Nottingham site.[21] Of the brands owned by the new division, six—Armstrong, James, New Hudson, Norman, Robin Hood and Sunbeam—had been discontinued within two years of the merger (*CTC Monthly Gazette* 80, 1961, no. 9: 211). By the time of the MMC report, only seven were still in use (MMC 1981: 8):

- the Raleigh brand itself
- Carlton (for high-specification sports bikes)
- Triumph and Sun (as off-season adult and sports brands, respectively, used to boost trade when sales were slow just after Christmas)
- BSA, Phillips, and Sunbeam, with specifications equivalent to Raleigh's other brands but sold only through mail-order catalogs.

The Chairman's statement in Raleigh's 1961 Directors' Report set the tone of the next two decades with regard to prospects for the workforce:

"The company is going to be engaged in a struggle over the next few years that will tax its strength and resources to the limit and we can only hope to emerge successful if we enter it stripped for action. If we carry surplus weight, whether in terms of labor or productive capacity, we shall fail and that can only have the most unhappy consequences for all those who earn their livelihood in this great company. . . ."[22] By the 1990s these predictions had been largely confirmed.

The year that followed the 1961 report saw large orders and double shifts, owing to tariff reductions in the United States.[23] However, by 1964 a drop in the US market had reversed this. An attempt to make 25 tool-room workers redundant resulted in a seventeen-week strike that affected 1600 more workers down the line.[24] This dispute was the first manifestation of Raleigh's legacy from the 1950s and earlier: both management and workforce were still geared toward an expanding market that no longer existed.

When management began to restructure the company to take account of shrinking sales, the Raleigh workers resisted changes—especially changes to the pay structure, which the company attempted to institute in 1967 after a three-year job-evaluation program. The New Wages Policy (NWP), negotiated between management and trade union leaders, aimed to replace piece rates with day rates, reducing wage differentiation across the company through fewer incentive payments (Cooper 1993: 29–30). The same year that this proposal was published saw a new outbreak of industrial strife, though, and the NWP was finally rejected by a workforce ballot in 1969.[25] It was imposed by management anyway, causing yet more strikes (ibid.: 30).

It was conflict some years later, though, that led to the introduction of new production technology and new shop-floor organizational structures intended to reduce Raleigh's vulnerability to industrial disputes. Dissatisfaction with wages led first (in late 1974) to a strike of several weeks' duration (ibid.: 31), then (in late 1977) to a costly series of longer strikes. After a number of weekly one-day strikes, the entire workforce walked out in mid November 1977, not returning until just before Christmas. Raleigh claimed this strike cost £3 million in lost production during the industry's peak sales period. The strike ended when the worn-out workforce accepted the same terms that shop stewards had earlier rejected: a 10 percent increase in wages, a 4 percent productivity deal with fringe benefits, removal of overtime and shift payment deductions, and better health-and-safety conditions.[26] Unlike the shop stewards, trade union officials did recognize the long-term threat this strike posed to the future of the industry (ibid.), a threat that was close to being realized several times in the next decade.

Almost immediately, TI began to discuss a modernization program for Raleigh that was first set in motion in 1979 after being piloted at the company's Gazelle factory in the Netherlands the previous year.[27] This program focused on the replacement of the existing conveyor system with short-line assembly tracks staffed by cells of about six workers. (By the 1990s, each cell included from eighteen to twenty workers.) The new system gave workers direct responsibility for quality control and was presented as being both more efficient and beneficial for employees in terms of worker autonomy and psychological well-being.[28]

Despite these changes in outlook, TI's approach was still, according to Frank Ellis, Works Personnel Manager in the 1990s, based on a view that spending money to increase production would reverse the company's losses.[29] During the early 1980s, TI invested £6 million in a computerized stock-control system and in new automatic and robot technology for welding, painting, and wheel production.[30] The introduction of this equipment caused serious supply problems that confirmed criticisms of incompetence made against the Raleigh management during the 1977 dispute.[31] These criticisms related back to TI's almost complete lack of investment in its Cycle Division between the mid 1960s and the late 1970s. The modernization program was long overdue. However, it coincided with the sharp decline in trade that had resulted from the loss of Raleigh's main overseas markets. Reduced orders thus combined with investment in new technology to cut Raleigh's profits, output, and workforce drastically. The workforce was reduced from about 9,000 workers producing 2 million bikes in 1978 to 4,000 producing 1.5 million in 1982.[32] By 1986, there were only 1,800 workers, producing only 1 million bikes.[33] After 200 redundancies in 1995 and 50 more in 1999, there were at most 500 core workers and 300 seasonal workers. In the mid 1990s approximately 750,000 bicycles were being produced per year; in 2000 (a bad year overall for cycle sales) fewer than 500,000 were produced.[34]

During the mid 1980s, TI substantially altered its approach to Raleigh's decline. Rather than prioritize expansion, the company began to adopt a range of practices linked to the perceived efficiencies of Japanese organizational practices, following the report of a task force of consultants and TI management commissioned in 1985.[35] This report's recommendations, which were adopted with little or no question, were based on a more extensive shift toward "cellular manufacture," which focused production on products rather than processes. Instead of the mass movement of products through processes (pressing, painting, assembly, and so on), production was restructured around cells of workers concerned with taking individual products through the various processes.[36] In addition, the company's production-floor space was cut to just 10 acres of its 64-acre site, eliminating one of the last remaining inefficiencies of the 1950s, whereby incoming materials could "travel several miles in the production process crossing two roads, a river, and a mainline railway."[37] In the 1990s—with Raleigh no longer owned by TI—the works was again "consolidated," contracting the factory-floor space into an area that no longer required crossing these barriers. Instead, it was reduced to occupy a smaller block of land situated between two suburban boulevards that it had previously straddled. One consequence of cutting factory-floor space was that there was no longer sufficient room to carry out all manufacturing processes—there was therefore a need to reduce the amount of production carried out in the factory. The solution was to stop producing front forks and instead to import them. (This was the reason for the 1995 redundancies.) In May 1999 it was announced that the company would cease building frames and now would import them. Thus, the factory's work was reconfigured to involve only the assembly of bought-in parts (*Bicycle Business* 1, July-August 1999). This allowed a profitable sale of the site to Nottingham University. Bicycle assembly is scheduled to move to a £14 million, 14-acre factory in the Nottingham suburb of Bulwell in 2003.

Raleigh after TI

The changes at Raleigh have affected the immediate area significantly. A walk around the former boundaries of the site provides evidence both of the grandeur of the 1930s and of the subsequent decline. The impressive architecture of Raleigh's former prosperity is juxtaposed with the decay of broken windows in now unused buildings. About half of the former site was sold in the 1980s for a housing development. Raleigh's former offices on Lenton Boulevard have been bought by the local council and leased to community groups. The company's former buildings on the far side of Triumph Road were bought by Nottingham University in the 1990s. The imposing modernist frontage of the 1952 factory on Triumph Road still dominates the scene, but the extent to which cars have replaced bicycles as a mode of transportation for the employees is shocking—especially to anyone who has seen the opening sequence of the film *Saturday Night and Sunday Morning*.

Figure 5.4
An aerial view of the Raleigh factory in the 1950s. Source: Raleigh Industries 1952b.

Figure 5.5
A 1993 photo (by the author) of Raleigh's factory on Triumph Road in Nottingham.

At Raleigh, cellular manufacture has been associated with reduced inventories of parts (i.e., just-in-time production), with the complete elimination of William Raven's conveyor system in favor of short tracks, and (until the company abandoned frame building) with the routine use of computerization, robots, and lasers for cutting tubes, welding frames, truing wheels, and so on. These processes were accompanied by an increased standard batch size of 500 bicycles instead of the erratic batches (often as small as 10) allowed previously. The greater ability to respond to market demand has been matched by the more sophisticated use of information about markets and competitors. Raleigh continually assesses the cycle market, performing some market research itself and contracting other research from specialized firms.[38]

The increased efficiency that quickly followed Raleigh's adoption of new production and marketing approaches enabled TI to sell Raleigh as a going concern, albeit at a loss, to Derby International, a conglomerate of American cycle-enthusiast industrialists and financiers set up especially for the purpose, in 1987. In its first seven years alone, Derby invested more than £14 million in Raleigh, expanding the use of robot technology and advanced laser welding equipment (*Independent*, March 7, 1994: 25).

Under Derby, Raleigh quickly began to make profits after several years of multi-million-pound losses. The company attributed the improvement to reductions in overhead and to adjustments of working hours made to suit seasonal sales patterns.[39] (The latter was another practice drawn from the repertoire of flexible specialization.) Raleigh now has a core workforce whose annual hours are divided up to suit the company's seasonal requirements. After Christmas, workers earn the same averaged-out wage for a 32-hour week as they do in the 45-hour week peak season before Christmas, when they are supplemented by temporary workers with less job security. How happy the core workers are with this arrangement depends on how it is operationalized in their particular department—some are much less happy than others about the notion that they owe "hours" to the company until the peak period begins, a situation that prevents them being able then to claim overtime payments.[40]

At the level of management, the increased productivity under Derby was seen to be the result of greater flexibility compared with the rigidities of TI. For Yvonne Rix, Marketing Director in the mid 1990s, the sale to Derby was "like being let off a lead . . . because the Derby philosophy was—well, first of all they got rid of about four or five key people, and I moved from being Product Planning Manager to Marketing Director Designate instantly, and there were only five people on the board, and it was wonderful, and it was 'just make these decisions yourself,' you don't have to ask anyone, you don't have to fill a form out in triplicate, just go ahead and run the company. And I don't mind if you make mistakes, but don't dig holes for anyone. . . . He does mind, but he says, his belief is if you don't make mistakes you're not doing anything. It's settled down a little bit, because after the first euphoria there's been more people appointed and a slightly more rigid regime. It's very good." As this quotation shows, the new practices at Raleigh display many of the main features of flexible specialization or post-Fordism, and these changes are portrayed by management as being primarily of benefit to the workers. "Treating people like morons or automata, just doing one thing day after day, is ridiculous," Yvonne Rix noted. Some of Raleigh's trade union representatives seem to regard the changes as for the better, given the precariousness of the British cycle industry. Cellular production was applauded by one unionist for allowing workers to take breaks with their colleagues' cooperation—not such a minor point in comparison with the rigidities on the line at Ford during the early to mid twentieth century or even at "post-Fordist" General Motors and Nissan (Beynon 1975; Parker and Slaughter 1990; Garrahan and Stewart 1993).

Another typical feature of flexible specialization that has figured in Raleigh's new approach is Total Quality Management (TQM), adopted by the new regime in the late 1980s. TQM is widely regarded with suspicion among critical analysts of the changing workplace (see Hobson 1992). The rationale behind its adoption at Raleigh (under the flag of "Success Through Quality") was presented both internally and externally in terms of a "culture change" that prioritized improving communications and improving the company's competitive position within the industry.[41] However, its introduction and its progress were regarded ambivalently by some in management as well as by union representatives. Most tellingly, the apparent elimination of some of the hierarchy of the shop floor, brought about by replacing supervisors and "leading hands" with "team leaders," was applauded by management personnel as a significant improvement but was seen by union representatives as old wine in new bottles. Nevertheless, despite the widely held sense that management communication had not lived up to the initial promise of TQM, when I visited Raleigh in 1997 and 1998 workers and management were feeling optimistic about the appointment of a dynamic new Managing Director, Mark Todd, whose commitment to new management models was believed to be more sincere than that of his predecessors. Todd's two-year career in the company (he resigned in 1999) was seen by the cycle industry press to have marked an era of radical change for Raleigh—a "root and branch overhaul" (*Bicycle Business* 2, September 1999: 4).

While the changes in the organizational structure at Raleigh have provided benefits for a wide range of actors at a time of great stress on the British bicycle industry, their clearest benefits have been for management, for whom they provided a means of addressing the problems that arose out of the conflicts of the 1960s and the 1970s. Giving workers responsibility for quality control and for production levels allowed Raleigh to move beyond the seemingly irresolvable conflicts that had been damaging it for nearly 20 years. Furthermore, putting workers into cells, so that they have mutual responsibilities, provided a deterrent to actions that might threaten the jobs of one's colleagues and also reduced the chances of the whole factory's being forced to shut down by an industrial action in one section. Now that workers are trained to do a number of different jobs across the whole range of bicycle manufacture, a strike at one end of the production process can no longer lead to shortages of work for those down the line. The motivation behind the changes at Raleigh as far back as the late 1970s can be seen, then, to have been aimed at reducing the ability of labor to affect trade as much as at reversing the decline caused by mismanagement and lost markets or at improving organizational communication. In this, it bears much in common with the pursuit of wheel-building machines intended to displace the recalcitrant Tivey in the interwar years, and thus it shows that the use of technology to displace labor—especially troublesome labor—has remained a consistent feature of the bicycle industry over the years.

Does this consistency mean there is little difference between the Raleigh of the early twentieth century and the Raleigh of the early twenty-first century? The changing management rhetorics traced in this chapter show that such is not the case. Furthermore, though the severe diminishing of a once triumphant company is in many ways unfortunate, and though the termination of production at Raleigh is a sad end to a crucial component of Britain's industrial heritage, I would argue that in several respects the present-day situation at Raleigh is far healthier than what once prevailed. First, the company's situation is transparent to all organizational groupings—no one, whether management or labor, holds any illusions about the cycle market or the company's productive capacity; all are equally aware of its fragility and hence of the need for cooperative visions of how the situation might be improved. Second, the industry is no longer based on a relationship with former colonies which was problematic both because of the ethical dimensions of the power relations involved but also in terms of how wise it is for an industry to be reliant for 60 percent of its output on such an unstable relationship (as was proved in 1979). Finally, as I will show, abandoning mass production in favor of more flexible technologies and management strategies has put the industry closer in touch with consumers and has allowed it to tailor its products more closely to users' "needs."

This is not to say that Raleigh's situation since the late 1980s lends more than partial credence to either the theorists or the critics of the phenomenon known as "post-Fordism" or as "flexible specialization." Just as with Fordism, it is hard to pin down what constitutes "post-Fordism," or how well such a label can be usefully applied across a range of organizations and industries (Williams et al. 1987; Jessop 1991a). This makes "post-Fordism" and related terms problematic to use without careful definition and elaboration. Thus, it is not surprising that, just as Raleigh's mid-twentieth-century practices represented something more complex than Fordism, its situation since the late 1970s been far more complex than any ideal-typical post-Fordist scenario. Nevertheless, certain features of this scenario—especially the notions of globalization and flexibility—have been crucial elements in the changing faces of bicycle production, of the cycling culture, and of the bicycle market.

6

The Global Flexibilization of the Bicycle[1]

The disintegration and the reconfiguration of the components that made up the sociotechnical frame of the mass bicycle formed one part of the transition to the new frame that became established by the 1990s: the globally flexible bicycle. The disappearance of many of the rigidities of mass production and its mass domestic and overseas markets left the way clear for new forms of production and industrial organization to emerge, linked to new supply chains, new markets, and new consumer cultures. These were accompanied by the appearance of new products and new cultural values associated with the bicycle and with cycling. In this chapter I will examine these emerging elements, which have transformed the production and consumption of bicycles and the culture of cycling.

Fragmented Production and a Changing Market

Although the merger with TI in 1960 gave Britain's new cycle conglomerate approximately 80 percent of the industry's yearly output of 2.25 million bicycles, by the 1970s both Britain's output and Raleigh's share of it were beginning to drop. Since then, there has been a radical transformation of the makeup of both the industry and the market, not just in Britain but worldwide. Domestic production, for both home sales and export, began to drop in the 1960s, reaching a relatively stable position by the 1980s. The only significant peaks in recent years have been during the booms of 1983–84 and 1989–90, responding respectively to the demand for BMX bikes and the demand for mountain bikes.

However, domestic sales have been rising steadily since the early 1970s, linked to a sharp rise in imports from practically nothing in the late 1960s to more than 50 percent of sales by the early 1990s. (See table 6.1.) Most significantly, the main source of these imports is no longer Europe; it is now the Far East. Imports from the Far East jumped from 1.9 percent to

Table 6.1
Data on British cycle industry (in thousands), 1960s–1990s, amalgamated from the following sources: MMC 1981: 4; Bicycle Association 1993: 6; TI plc Interim Statement 1983 (LSL qL62.9, "TI Annual Reports" box); *Financial Times,* March 7, 1986, *Bicycle Business* trade figures.

	Total UK			
	Production[a]	Imports	Raleigh's UK sales	Total UK Sales[b]
1965	1,630	—	—	728
1967	1,490	5 (0.85%)[c]	—	590
1970	1,648	27 (4%)	375 (59%)	641
1975	1,952	273 (25%)	595 (54%)	1,104
1980	1,746	563 (36%)	628 (40%)	1,563
1983	1,550	840 (39%)	1,100? (>50%)	2,150
1984	1,470	840 (41%)	—	2,050
1985	1,244	540 (36%)	800(?) (45%)	1,514
1987	1,176	1,023 (51%)	—	2,000
1988	1,065	1,305 (59%)	—	2,200
1989	1,387	1,593 (57%)	—	2,800
1990	1,275	1,771 (63%)	1,000 (36)	2,800
1991	1,134	1,589 (63%)	750 (30)	2,525
1992	1,180	1,700 (64%)	750 (28)	2,650
1993	1,000	—	—	2,400
1994	1,150	—	—	2,200
1995	1,100	—	—	—
1996	—	—	—	2,600
1997	—	—	—	2,400
1998	—	—	—	2,100

a. Includes production for domestic sale and export.
b. These numbers may represent deliveries to retailers rather than actual sales, especially in the peak years of 1983–84 and 1989–90.
c. i.e., percentage of domestic sales

9.9 percent between 1979 and 1980 (MMC 1981: 6). By 1988, imports from Taiwan, Hong Kong, China, Thailand, and Indonesia together accounted for 32 percent of UK sales,[2] while in the period from January to August of 1992 Far Eastern bicycles (including bikes from Malaysia and Japan as well as bikes from the countries already mentioned) accounted for 74 percent of imports.[3]

Alongside these changes in production and sales, there has been a renewed fragmentation of the industry that has reversed the growing convergence of production that occurred during the middle of the twentieth century, a situation that appears at one level to be following the "disorganization of capital" described by Lash and Urry (1987). Raleigh no longer forms the bulk of the British cycle industry, its market share having been reduced by a range of competitors, and the industry as a whole is in something of a state of flux in terms of company ownership and market stability. Nevertheless, cycle production is not as disorganized as it might look from Lash and Urry's perspective. In particular, the great diversity of cycle brands masks a concentration of ownership and production, and a complex ordering of the patterns of trading relations, which indicate a quite structured process of globalized organization (Hirst and Thompson 1996; Gereffi 1994) in which niche marketing and branding—and hence the *appearance* of disorganization—are strong features (Abercrombie et al. 1990).

Precise production and sales figures for the industry are uncertain, and different sources contradict one another. (For example, the combined domestic sales for Raleigh and all imports in table 6.1 seem to leave too little space for other non-imported brands.) Nevertheless, a roughly agreed-upon breakdown of the industry in the 1990s gives Raleigh 30–35 percent of the home market, depending on whether one is referring to the numbers or the value of sales. Raleigh's biggest competitor for some years have been the Townsend, British Eagle, and Falcon brands, owned by the Casket Group until 1995, when losses in other areas of the business led to a merger with the Tandem Group (*Financial Times*, May 30, 1996 and January 8, 1997). This merger itself was not entirely stable at the time of writing, having lost some of its nearly 30 percent market share to newer players in the industry.[4] After Casket, the only companies with a significant market share—both approximately 10 percent—are Universal Cycles (producers of bicycles mainly for the very lowest end of the market) and Moore Large (an importer whose brands have included Emmelle, Scott, and Diamond Back).[5] Other smaller companies, each with a small share of the market, include the Taiwanese company Giant, the American companies Trek and Cannondale, and the British companies Dawes and Muddy Fox. Few of these companies are committed to production in the United Kingdom. Raleigh was a proud exception until 1999, when it became an assembler of bought-in parts. Other companies continue to mix home production with importing. For example, Townsend opened a new factory in 1994 (*Cycling Today* July 1994: 10), and there are plans to eventually begin building frames in Wales at a Taiwanese-owned assembly plant that was opened in 2000 (*Bicycle Business* no. 9, 2000: 20–21).

Such expansion has been facilitated by the general growth in cycle sales since the 1980s. This growth, which revived the British industry after its decline in the 1970s, resulted from a number of related cultural trends. Cycling has been central to new concerns with leisure and health. It is seen as a solution to the perceived environmental crisis and also as a means of transportation that could solve problems of traffic congestion. Such developments have been underpinned in the United Kingdom by a variety of policy changes having to do with cycling's health, economic, and planning benefits (BMA 1992; CTC 1993; Department of Transport 1996b; DETR 1998). The overall result has been a growth in the purchase of bicycles, especially for leisure use. About 15 percent of adults in the United Kingdom owned a bicycle in 1976, about 30 percent do today, accounting for 2.3 percent of journeys (but as much as 28 percent in some cities).[6] This increase, and the new meanings that are attached to bicycles, have been helped to a strong degree by changes in bicycle technology. In particular, mountain bikes are perceived as safer and more enjoyable to ride because of their sturdy design and upright riding position, and because of the improved braking and gearing technologies.

A new sociotechnical frame of the globally flexible bicycle has, then, emerged since the 1980s, centered on several interlinking features: new production methods and new ways of organizing the industry; a new configuring of the industry on a global scale; new products; new ways in which consumers perceive and use bicycles. In this chapter I will examine the global flexibilization of the bicycle and the cycle industry and the role of mountain bikes in this process. I will also explore the cultural contradictions that emerge with a new artifact which is at once pollution-free yet environmentally questionable in its design and use, which democratizes accessibility yet is technologically opaque, and which is a potential solution to transportation problems yet in many ways depends on and increases the use of automobiles.

Globalization and Flexibility: A Second Industrial Divide?

To begin this exploration, I want to return to Piore and Sabel's (1984) thesis of the "industrial divide"—the apparent choice made by both industries and regions between mass production and flexible specialization as a main industrial strategy. The global flexibilization of the bicycle industry in the 1980s and the 1990s raises serious questions about the credibility of this model. For example, a UK producer that contrasts with the mass approach of Raleigh and its larger competitors is Dawes, which makes about 30,000 bicycles a year in a new factory (opened in 1993) in Birmingham. Unlike the larger producers, Dawes at this stage was using very little automation, except in a few areas such as its automated lacquering line. Only the lower-price models were built on an assembly track, and Dawes had not invested in CAD/CAM (computer-aided design/computer-aided manufacturing) systems, leaving design to be done "on the drawing board" or (more often) in the designer's head. The company adopted computerization only to administer its dealer network.[7]

The rationale behind Dawes's approach has been the company's emphasis on flexibility and its desire to maintain a wide and flexible range of products. In the terminology of Piore and Sabel, Dawes has a form of flexible specialization, whereas Raleigh and Townsend have something closely resembling mass production—a duality that has existed in the cycle industry throughout its history.[8] (See especially Hudson 1960.) Dawes was founded in 1926, when mass production in the industry was building toward its peak. Yet Dawes has never chosen the path of mass production.

With the growing diversity of cycle firms, the role of flexible specialization in the industry appears to be rising. It is questionable, though, whether the evidence supports Piore and Sabel's claim of the mid 1980s that industry and regions had come to a "second industrial divide" between mass production and flexible specialization and needed to choose either to embrace the latter or to abandon it in favor of a renewed commitment to mass production and Keynesianism (Piore and Sabel 1984). Hudson (1960) provides support for critiques of this thesis (e.g. Williams et al. 1987), making it clear that even in the earlier mass-production period—at the time of Piore and Sabel's "first industrial divide"—the work of mass producers and craft producers in the cycle industry was complementary rather than competitive. Large-scale producers depended on flexible specialists to develop and test innovations (in terms of both workability and market interest) without the risks entailed by setting up vast equipment to produce what might prove to be badly designed or unpopular machines. Smaller producers earned a steady income from specialized demand to which larger concerns could not cater, and they also benefited from the increased popularity of cycling caused by lower prices.

In the 1990s, as a result of the restructuring of the industry, even the distinction between mass producers and flexible specialists in the cycle industry became less clear than ever. This has been compounded by the fact that since the 1970s the globalization of the industry has thoroughly transformed its structure on an international scale too (Sutcliffe 1991: 38–39). This has meant that it is increasingly difficult to speak with certainty of a *British* bicycle industry as the discrete entity it was until the 1970s. Although there are British manufacturers of bicycles, bicycle components, and accessories, these are all tied into a variety of global networks of relationships among manufacturers, suppliers, subcontractors, and consumers. For example, Raleigh itself remains a British company in terms of location, staffing, and heritage, but from 1987 to 2001 it was owned (along with many of its former worldwide subsidiaries) by Derby Cycle Corporation, the world's largest bicycle group (Garnett 1989; *Financial Times*, May 16, 1998). In the 1990s, Dawes (owned for several years by the Dutch cycle group Atag) underwent a management buyout before being bought by the Tandem group in summer 2001. When Muddy Fox, the pioneering British mountain bike company, went into receivership in March 1992, it was bought jointly by the British property and engineering company Sitac and TI Cycles of India. The latter is the former Indian subsidiary of Raleigh, and it invested in Muddy Fox as a means of gaining access to the British market for its own cheaply produced bicycles (sold under the Silver Fox brand)—an ironic reversal of its former colonial relationship with Raleigh.[9]

More importantly, though, bicycles *marketed* by a small "flexibly specialist" company based in Britain or in the United States will probably have been designed and built in Taiwan by a huge company such as Giant, which produces almost 1.6 million bicycles a year under a wide variety of brand names alongside its own brand (*Financial Times*, January 31, 1989; Giant catalog 1992: 2; Sutcliffe 1991: 39; Hemsworth 1992). Taiwan took over this role in worldwide cycle supply from Japan during the 1980s. While there are signs of a further shift to cheaper labor in China and the Pacific Rim, the apparently minimal impact of the 1990s' Asian financial crisis on Taiwanese and Japanese firms has left the Taiwanese cycle industry in a strong position. Rather than lose market share to competitor regions, Taiwanese firms have opened their own overseas production centers. Meanwhile, Japan retains its prime role in the supply of cycle components, again drawing on cheap labor in other parts of the Far East. As a consequence, most bikes produced in Taiwan for British cycle firms are equipped with components produced by Shimano, a company that dominates the Western market for components.

This overall situation raises serious questions about the validity of the distinction between mass production and flexible specialization, both of which might be involved at different points and in different global locations in the construction of a single bicycle.

This transformation in the global structure of cycle production can be understood in relation to Gereffi's (1994) concept of *global commodity chains*. Gereffi identifies two different kinds of global commodity chain: those that are producer-driven and those that are buyer-driven. In common with other critical analysts of transnational economic change (e.g. Hirst and Thompson 1996), he identifies not a blanket process of "globalization" but rather a shift toward specific patterns in the internationalization of economic activity. The two types of global commodity chain he identifies represent the two ways in which this internationalization has been structured: producer-driven chains are dominated by transnational manufacturing firms that maintain centralized control of global production facilities, while buyer-driven chains are led by retailers and merchandisers in developed countries who use their economic power to control global networks of suppliers. Although this distinction can be regarded as mapping onto the concepts of mass production and flexible specialization, there are two important contrasts with Piore and Sabel's approach. First, there is no sense of a need to choose between the two type of chain, both of which can be identified in different contexts. Second, both kinds of chain are seen crucially as global, in contrast to the regional and national focus of Piore and Sabel (Gereffi 1994: 218). Nevertheless, the contrast between the two kinds of chain highlights the different kinds of product upon which discussions of mass production and of flexible specialization focus. Gereffi sees producer-driven global commodity chains as concentrated in traditional heavy industry sectors, including automobiles and aerospace, as well as in newer sectors such as computer engineering (ibid.: 216). In contrast, buyer-driven chains are the norm for clothing and sports equipment, the products that Piore and Sabel cite in their account of flexible specialization.

Gereffi's analysis clearly has resonance for what has been happening in the cycle industry since the 1980s. As a part of the development of global commodity chains, Gereffi describes *triangle manufacturing*, where firms in the newly industrialized countries of the Pacific Rim act as trading company subcontractors for Western manufacturers, illustrated in the way a Muddy Fox designer explained the procedure of designing and specifying a mountain bike[10]:

A. Tubing is selected, usually like the Tange series [a Japanese brand]. Many companies use no-name tubes supplied by factories, often of dubious quality.

B. Shimano supply "groupsets" [of components], through the Far East factories. Many companies do not use the full "group" to save money, just utilizing the brakes + transmission.
C. Frames can be selected off the shelf [e.g., figure 6.1], though most detail their own designs—
D. So the factory will submit a drawing like [figure 6.2]. It is pretty simple, but suffices for most manufacturers. Better companies will go into more detail.
E. Ancillary components are selected from a vast catalogue, the Taiwanese Bicycle Guide.

Such procedures are common among newer entrants to the bicycle industry. For such companies, "design" is often a matter of picking and mixing pre-existing elements rather than originating new ones. This form of "remote control manufacturing"[11] reduces cycle "manufacturers" in many cases to "little more than marketing companies with just an office and a phone" (*Ethical Consumer* 6, 1990, February-March: 19). Muddy Fox was at one point the exemplar of such an approach, having popularized mountain bikes in Britain through high-profile marketing that, for some, obscured the company's lack of design innovation, which was instead left to its Japanese and (later) Taiwanese trading partners.

Muddy Fox's early success is commonly attributed to its strategic marketing of a specific lifestyle rather than to the quality of its bikes (Rayner 1992). Its most popular model, the Courier, has been described as "a victory for sourcing and pricing" rather than one for design (Hilton Holloway, *Bicycle*, June 1992: 16). What sold it, for many commentators, were "color, style, and status" (Rayner 1992). A direct concern with technical matters developed at Muddy Fox only in the late 1980s, around the time of the mountain bike boom, when the company began to employ its own designers for the first time. It didn't promote a racing team until the mid 1990s, despite having recognized some time earlier that technological credibility depends on this.[12] An amateur mountain bike racer told me that he regarded Muddy Fox's mountain bikes as ideal for people who don't race. Only after the company's revival under new ownership did it begin to gain credibility, notably through experiments with suspension.

The flexibly specialist approach at Muddy Fox involves several characteristics highlighted by Piore and Sabel (including a high degree of innovation and responsiveness to the market), but is just one element of a more complex set of practices and relationships within the industry. Other firms display different configurations of innovation, technology, and organizational and cultural practices. In contrast to Muddy Fox's

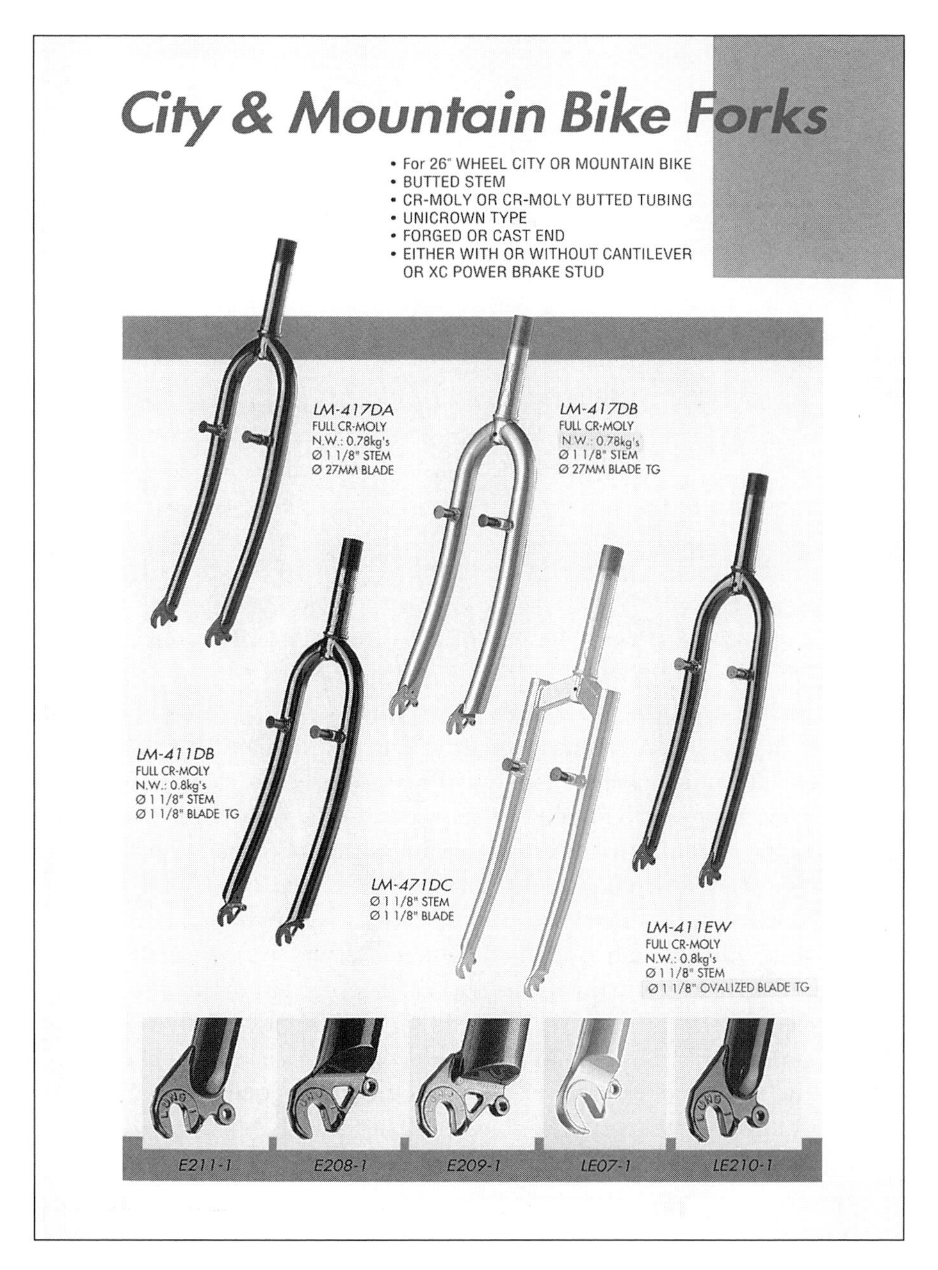

Figure 6.1
Front forks in the catalog of the Lung I Machinery Works, located in Taiwan.

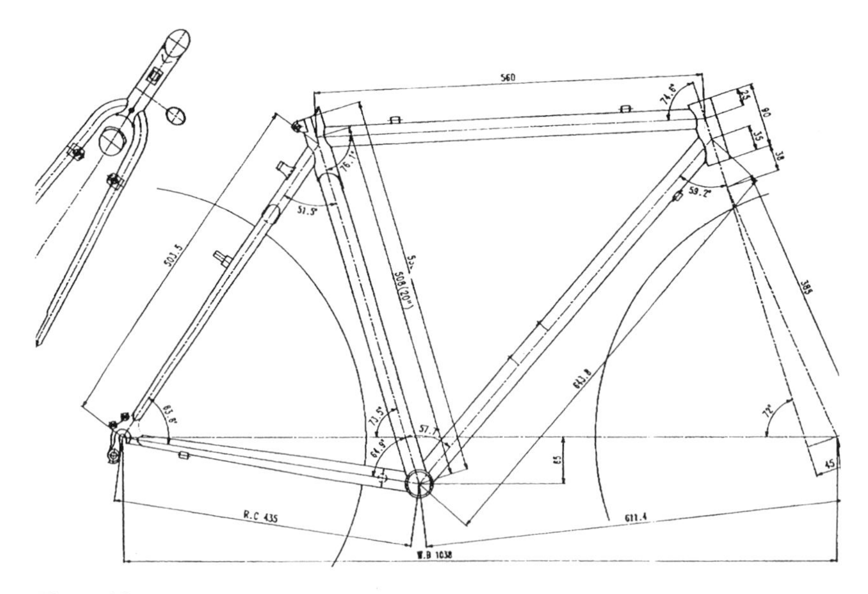

Figure 6.2
A diagram showing specifications for a bicycle frame. Source: letter from Hilton Holloway.

appeal to style, many companies specializing in mountain bikes concentrate on high-quality sports models in the higher price ranges, often manufacturing a very small number of machines. In the early 1990s, Orange, a company based in Cumbria, was producing approximately 1,500 bicycles a year at prices ranging from £600 to more than £2,000. Zinn, based in Birmingham, produced only 150 machines a year—which, unusually, they built, finished, and assembled. Between specialist producers such as these (which catered to the niche market for top-quality mountain bikes) and medium-size and large producers such as Dawes and Raleigh are companies such as Orbit and Pashley, which cover a variety of niche markets, including those for conventional touring and racing bikes, traditional utility models, recumbents, and folding models.[13]

The complexity of flexible specialization in the industry is reflected in the 1990s marketing strategies of Raleigh, whose more expensive models have for many years been differentiated from the company's other ranges by the use of branding, and by building them (until recently) in a separate factory at the Nottingham site under the auspices of Raleigh's Special Products Division. Workers there were regarded as more skilled—and earned more—than the rest of the workforce. Raleigh

Figure 6.3
A mid-1980s Muddy Fox advertisement.

emphasized the craft basis of production in this factory, where each bicycle was assembled by a single individual, whose name and photograph were attached to the bike before delivery to retailers, in contrast to the mass-produced image of the main Raleigh brand. Products included high-specification mountain bikes and touring models, produced in far smaller numbers than the standard run of 500 in the main factory. They had separate catalogs and were promoted through the Team Raleigh mountain bike racing team. At one point, the Raleigh name and heron's head badge were entirely absent from these higher-specification models. The advanced production technology which was widespread throughout the factory by the time frame building was discontinued had generally appeared in the Special Products Division first—for example, plasma arc welding and robotic frame building. Titanium, aluminum, and metal matrix composites were often used.

With Raleigh's most recent restructuring, the role of the Special Products Division was taken over by Diamond Back, the mountain bike specialist subsidiary acquired in 1999, following several years of trying to earn back a reputation as an innovator among specialist consumers who derided the company's old-fashioned image. In 2001 the company decided to focus just on the mass market, downgrading the Diamond Back brand. Its efforts over the years at product differentiation highlight the rhetorical importance of branding in maintaining a distinction between mass production and flexible specialization (Abercrombie et al. 1990), even when they are practiced simultaneously by just one company.

How can the situation just described be accounted for using critical approaches in the sociology of technology? The situation at Raleigh and elsewhere indicates that the position of Piore and Sabel (1984)—that it is necessary for companies, states, or regions to choose between mass and flexible approaches to production—is an inaccurate and untenable account of how industry has changed since the heyday of mass production (Williams et al. 1987). In many industries, companies have been using branding to differentiate products for many decades, and this is certainly true of the cycle industry, for example with Raleigh's earlier Gazelle and Robin Hood brands. Indeed, the Special Products Division had its roots in the purchase, shortly before Raleigh's merger with TI, of the custom cycle builder Carlton Cycles, which even in the late 1950s Raleigh's management felt would improve the company's credibility (Bowden 1975: 99).

The differentiation achieved through the acquisition of specialist subsidiaries is not, then, a new development, and it embodies a number of contradictions that are present more generally throughout the cycle industry. First, it demonstrates a *continuity* across the eras commonly characterized as Fordist and post-Fordist (or between the sociotechnical frames of the mass and globally flexible bicycles), which are ideal-typically portrayed as discontinuous (although most commentators recognize that the disjunction is overplayed). Second, in common with the significance of Taiwanese factories and Japanese components to small British mountain bike companies, the location of Raleigh's island of flexible specialization within the geographical, managerial, and economic boundaries of mass production indicates a need to reassess the status of both categories. Greater flexibility in production offers clear economic advantages in allowing access to more diverse markets, and hence to increased profits, without needing to continually reinvest in new equipment. It also has great value to a company such as Raleigh (or indeed to the smaller manufacturers) in a way that is underplayed by the advocates of post-Fordism and flexible specialization. This value centers on the discursive advantage in being able to convince customers that a particular product is something special, set apart from the rest. Indeed, distinguishing flexible specialization from mass production on any grounds other than discursive ones is bound to be problematic—Williams et al. (1987: 414–417) point out how difficult it is to make credible distinctions between the two approaches on the basis of features that are commonly cited as decisive—the use of dedicated equipment, product differentiation, and length of production run.

Rather than rely on such questionable distinctions, it is perhaps easier to concur with Harvey (1989) in seeing recent economic, industrial, and cultural changes as elements of broader trends in capitalist growth and crisis. For Harvey, despite its specific historical location within the global economic changes of the 1970s and the 1980s, "flexible accumulation is still a form of capitalism," and consequently it is subject to the usual inconsistencies and contradictions between capital and labor and between innovation and accumulation discussed by Marx (ibid.: 179–180). The resulting overaccumulation, which is typical of capitalism, is "indicated by idle productive capacity, a glut of commodities and an excess of inventories, surplus money capital . . . and high unemployment" (ibid.: 180–181). Attempted solutions to the crises that inevitably (for Marx and for Harvey) result from this include the devaluation of commodities, productive capacity, money value and labor power; macroeconomic control through mechanisms such as Keynesianism or state Fordism; and the absorption of overaccumulation through temporal and spatial displacement. Here the temporal option "entails either a switch of resources from meeting current needs to exploring future uses, or an acceleration in turnover time . . . so that speed-up this year absorbs excess capacity from last year," while spatial displacement involves "the absorption of excess capital and labor in geographical expansion" (ibid.: 181–184).

Harvey's analysis provides a powerful tool for understanding changes in the cycle industry. His account of how overaccumulation of capital develops into crisis recalls the major crisis periods of cycle production: the 1890s, when the upper-class cycling craze led to speculation through the flotation of almost every major manufacturer; the resulting depression, which lasted several years; the buildup of capacity and the saturation of world markets from the 1930s to the 1950s, which forced the amalgamation of almost the entire industry under a single company (which then experienced a severe decline lasting 20 years); and the stagnation of the 1970s, which was due in part to unresolved earlier problems but which was exacerbated by a continuing commitment to expansion alongside the "external shocks" of oil crises and lost markets (which themselves highlighted the connection with the broader crisis of global capitalism).

Similarly, the solutions to crisis described by Harvey can be seen in the way cycle manufacturers have attempted to achieve recovery. Prices have been cut and cheaper models produced; at the same time, products have been differentiated so as not to be dependent on a single product line. Workers have been transferred from day rates to piece rates or (more recently) taken off piece rates and simultaneously induced to improve productivity through organizational change. Process innovations have been introduced to reduce unit costs. Workforces have been cut. Product innovations in the higher price range have been quickly brought into mass production. Factories have been reorganized to achieve more efficient handling of parts. New methods of stock control have been adopted. Production as a whole has been first aggregated, then dispersed and re-aggregated globally in ways almost too complex to describe adequately.

These events, identifiable in their different ways in each of the sociotechnical frames of the bicycle that I have been describing, highlight what has probably been a far greater degree of continuity than discontinuity in cycle production over the course of its history. Nevertheless, although the constant innovation characteristic of capitalism in general and cycle production specifically might be seen simply as evidence that "there are many ways to make a profit" (ibid.: 343), for Harvey this is an insufficient analysis. To regard what he terms flexible accumulation simply as "a jazzed-up version of the same old story of capitalism as usual" would be to treat capitalism "a-historically, as a non-dynamic mode of production, when all the evidence . . . is that capitalism is a constantly revolutionary force in world history, a force that perpetually re-shapes the world into new and often quite unexpected configurations . . . " (ibid.: 188).

Harvey's analysis falls short in its lack of attention to the ways in which "capitalism . . . re-shapes the world" not through some inherent metaphysical power but through the deeds of social actors who innovate economically, technologically, organizationally, and culturally. Harvey pays too little attention to the ways in which sociotechnical change is achieved in the day-to-day life of organizations, their workforces, their customers, and their suppliers. To redress this means asking whether the cycle industry's actors of the 1980s and the 1990s did anything substantially different from what their counterparts of the 1920s and the 1930s did. This question is bound up with whether and in what ways flexible production, globalization, and the fragmentation of bicycle markets differ qualitatively from the dynamic international capitalism that came before (Hirst and Thompson 1996). I will explore these questions further by extending the discussion from the changing dynamics of the cycle industry to the new products, production processes, and cycling cultures that have developed since the 1970s.

Mountain Bikes and the Shimano Paradox

The Cultural and Technological Construction of a New Artifact

Integral to the story of bicycles and cycling since the 1980s has been the rise of mountain bikes. Introduced into the United Kingdom in the early 1980s, mountain bikes quickly caught the imagination of the emerging "yuppie" culture, encouraged by the marketing approach of companies such as Muddy Fox. They were also an instant hit with the burgeoning cycle-courier fraternity, who saw this new stylish but sturdy bicycle as an ideal means of negotiating poorly maintained city streets. The cycle industry was more cautious, having had its fingers burned by the BMX boom of 1984, which left many manufacturers and retailers with excess stock. Consequently, mountain bikes took longer to reach the mainstream public; however, having done so in the late 1980s, they led to a new boom in the industry, and they quickly became the dominant design. In 1988 mountain bikes accounted for only 15 percent of the 2.2 million cycles sold in Britain; in 1990, 50–60 percent sold were mountain bikes, and these are seen to account for the increase in sales to 2.8 million that was sustained through 1989 and 1990 before dropping back in subsequent years (Bicycle Association 1991: 6). It is reasonable to assume that the popularity of mountain bikes has helped to keep the industry in business, especially in view of the fact that its rise coincided with Raleigh's transfer from TI to Derby.

Mountain bikes were not, though, invented by the industry to revive its ailing fortunes; rather, their emergence and their rising popularity provide an outstanding example of the serendipity of sociotechnical change, which cannot be predicted or easily directed toward particular desired objectives. Instead, a series of chance events and associations among actors happened to set off a trajectory that resulted in a new hobby, then a new sport, and eventually a revived industry centered on an object that has multiple meanings for a wide range of users.

Mountain bikes were "invented" in the 1970s.[14] Their technological predecessors first appeared in Marin County, California, in the early 1970s, as an integral part of the recreational activities of a group of "hard-core hippie bike bums" who had moved there from San Francisco "to live less frenetic, more laid-back lives" (Kelly and Crane 1990: 10). These people began to build downhill racing bikes, known as "clunkers," from frames and components that they found lying around in back yards. In particular, they used frames from the quintessential "newsboy" bike of the mid twentieth century, the Schwinn Excelsior. These frames were sturdy enough to withstand rough treatment from adults once they had

Figure 6.4
Gary Fisher's 1974 "clunker." Source: Gary Fisher catalog.

been treated to "the standard Marin County conversion," which entailed adding modern components such as "derailleur gearing systems . . . front and rear drum brakes, motorcycle brake levers, wide motocross handlebars, handlebar-mounted shift levers, and the biggest knobbly bicycle tires available mounted on heavy . . . steel rims" (Kelly and Crane 1990: 21). The primary task of such a machine was to survive what is now an infamous ride in mountain bike folklore: the "repack run," a steep drop on the slopes of Mount Tamalpais, losing 1,300 feet in less than 2 miles. Riders would be taken up the mountain in a truck and would then race down it on their clunkers. The ride got its name because by the time a rider reached the bottom all the grease in the original coaster brakes would have "turned to smoke" and the hub would have to be repacked with fresh grease (ibid.: 22).

The rapid growth of the original clunker group, and the subsequent spread of what became known as mountain bikes, has been credited to the cultural resonance these machines had for Americans who had grown up in the middle of the twentieth century. Patton (1991) writes about the decline in the United States of cycling as an adult activity from the early twentieth century until the appearance of fitness and environmental movements in the 1960s. He attributes the success of mountain bikes to their new role as "the long-missing American adult utility bike—a rugged, efficient, general-purpose bicycle" to replace the European racing bike (ibid.: 67).

Mountain bikes also, however, evoke memories of childhood, as is evident from the following quotation[15] celebrating the tenth anniversary of mountain bikes:

> "I grew up with an old Schwinn two-speed," recalls Lou Gonzalez, who promotes the subsport of Mountain Bike Polo. "We used to ride around on these forest trails in Illinois at breakneck speed, trying to kill each other.
>
> The first time I got a mountain bike on singletrack it reminded me of my youth. It brought back that special time when you didn't have a care in the world."

Mountain bikes are often said to evoke nostalgia in prosperous "baby boomers" (Patrick 1988). Nostalgia has been associated with mountain bikes in Britain too, especially with the growth in leisure cycling (Ruff and Mellors 1993).

The speed with which everyone who encountered clunkers in the late 1970s appears to have become converted to them (Kelly and Crane 1990; Ballantine 1988) resulted in a demand for custom-built bikes that improved on the Schwinn design. The person generally credited with building the first of these custom bikes was Joe Breeze. His design was improved on by Tom Ritchey, an established frame builder outside the original clunker group who set up a company called MountainBikes with two members of the group, Gary Fisher and Charles Kelly. This trio built, equipped, and marketed the first commercial mountain bikes, starting in 1979 (Kelly in Kelly and Crane 1990: 50–51). In 1981, another Californian cycle company, Specialized, had the Ritchey-Kelly-Fisher design manufactured in Japan as the first mass-produced mountain bike, the Stumpjumper. The appearance of the Stumpjumper at trade shows in 1982 was the general public's introduction to the mountain bike.

Shimano: Innovation and Control

Perhaps the most important factor in the rise of mountain bikes was the introduction by the component manufacturers Shimano and SunTour of special mountain bike components. In Kelly's words (ibid.: 55): "The availability of component groups was the last stage of assembling the infrastructure necessary for mass production, and from that time forward mountain bike production swung into high gear, maintaining for several years the highest growth curve in the bicycle industry. For better or worse, mountain bikes were no longer a garage industry. . . ." The fact that this was not mass production but globally flexible production does not diminish Shimano's significance to the progression of mountain bikes within the cycle industry and to the cycling culture. Shimano's ascendance to the status of the dominant component brand in Britain was contemporaneous with, and reinforced, the ascendance of the mountain bike. By 1987, Shimano was "widely acknowledged as cycling's technological leader" (*Freewheel Catalogue* 1987: 148). Through its innovative component design and its control over component supply, Shimano helped bring about the dominance of this new kind of bicycle while achieving dominance of the industry.

Shimano's rise to prominence, largely through an association with mountain bike innovations, was by no means serendipitous. Unlike many of the mountain bike manufacturers, Shimano was a long-established business whose activities offered clear evidence of strategic planning. It was founded in 1921 in Sakai City (near Osaka, the center of the Japanese cycle industry) to manufacture freewheels. In later decades it expanded to produce other components. Since 1970 it has also produced fishing equipment, which accounted in 1989 for 23 percent of its output while a further 6 percent was taken up by cold forging for auto

Figure 6.5
A 1992 advertisement for the Specialized Stumpjumper.

manufacturers and others. Medium-price and high-end components are produced at the company's factories in Sakai and Southern Japan; low-end products are made in Singapore, Malaysia, and Korea, where labor is cheaper. The company exports two-thirds of its output, most of which goes to Europe and the United States. In 1991, after-tax profits were reported to have risen 63 percent (*Financial Times*, July 12, 1991; other statistics from *Far Eastern Economic Review* 146, 1989, no. 50: 103–104; additional information from Espinoza 1992 and Thisdell 1990).

Shimano's strength has been due in large part to its astute recognition of the commercial potential of mountain bikes. The company was willing to invest in research and development of new components when the rest of the industry saw mountain bikes as a fad that would soon follow Choppers and BMX into near oblivion. In particular, Shimano has developed componentry that has made cycling far more "user friendly." Such Shimano innovations as indexed shifting, integrated gearing and braking systems, smoother transmission systems, and clipless pedaling were developed to reduce uncertainty about the performance, the reliability, and the efficiency of these functions.[16]

Shimano continually refines its products. Consider, for example, the revolutionary Shimano Index System (SIS), launched in 1987, which allowed the rider to change gears more quickly and easily than before. This system replaced "friction" gear shifting, which required the rider to move the gear lever up or down until the chain slid onto the next sprocket or chainwheel. It was easy to miss the right spot, thus jumping gears or even coming out of gear altogether while the chain floundered between sprockets. Indexed shifting uses preset staged positions to move the chain exactly onto the next sprocket, "clicking" into the right gear rather than leaving the rider to guess. SIS was supplemented with a succession of new kinds of shifters—first handlebar-mounted thumb shifters, then a more sophisticated dual-pushbutton system incorporating a ratchet mechanism adapted from the company's fishing line technology, then a system of thumb-and-forefinger-operated levers. Within 5 or 6 years, changing gears was transformed from guesswork and "feeling" to the certainty of pushbutton control, supplemented by display of the gear number. Similar innovations were taking place at the same time in other areas of bicycle componentry, including chains, freewheels, sprockets, and brakes.

The attention (and R&D investment) spent by Shimano on making these functions more straightforward and unproblematic for inexperienced cyclists is seen by many as a major factor in the growth of cycling

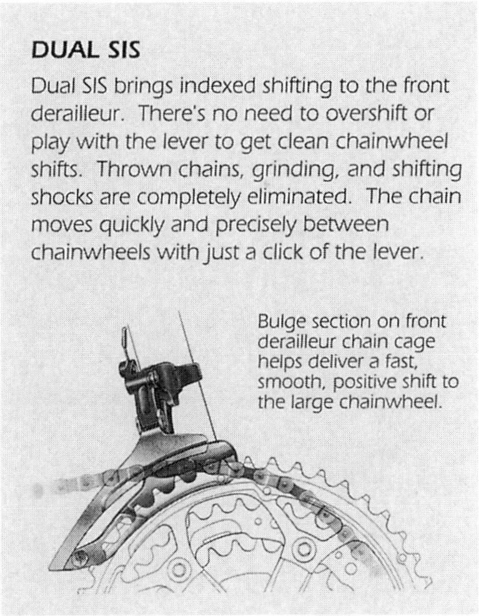

DUAL SIS

Dual SIS brings indexed shifting to the front derailleur. There's no need to overshift or play with the lever to get clean chainwheel shifts. Thrown chains, grinding, and shifting shocks are completely eliminated. The chain moves quickly and precisely between chainwheels with just a click of the lever.

Bulge section on front derailleur chain cage helps deliver a fast, smooth, positive shift to the large chainwheel.

Figure 6.6
Indexed shifters. Source: 1994 Shimano catalog.

during the 1980s. This view has been enhanced by the widespread perception that Shimano's innovations generally work very well and quickly filter down the price scale. As a result of this commitment to improving cycle technology, alongside the company's early support for mountain bikes, many mountain bike innovators and enthusiasts have been loyal to Shimano despite some ambivalence.

The ambivalence centers on two aspects of Shimano's relations with its customers: its perceived control over the manufacturers it supplies and the dependence it is seen as endeavoring to create in its end users. Bicycle manufacturers are almost unanimously critical of how Shimano treats them and the industry. Manufacturers are said to be pressured, through volume discounts, into buying complete Shimano groupsets rather than only certain parts. Firms have complained that Shimano refuses to supply product samples before release, and there are suspicions that the company has deliberately introduced inferior products in order to then enhance its reputation with subsequent refinements (*Mountain Bike Action* 6, 1991, no. 8: 94–108).

Most controversially, the company's innovatory approach has meant that, since the late 1980s, every year has seen radical changes in the company's product range. Table 6.2 shows how Shimano restructured its

Table 6.2
Shimano groupset series.

	1991	1992	1993	1994	1995	2001
Top of range		XTR	XTR	XTR	XTR	Airlines (downhill racing) XTR (professional racing)
Top-class racing	Deore XT Deore DX Deore LX	Deore XT Deore DX Deore LX	Deore XT Deore DX Deore LX	Deore XT Deore LX	Deore XT Deore LX	Deore XT Deore LX Deore
Off-road	Exage 500 LX Exage 400 LX Exage 300 LX	Exage 500 LX Exage 400 LX Exage 300 LX	Exage ES Exage LT	STX SE STX Alivio	STX RC STX Alivio	Alivio
Leisure			Altus A10 Altus A20	MJ Altus C50	Acera-X	Acera
Hybrid leisure			Altus C10 Altus C20		Altus C90	Altus
Budget	200GS 100GS	200GS 100GS 70GS	Tourney	Tourney 30 Tourney 20	Tourney groupsets	Tourney

range of groupsets between 1991 and 1995, set against the 2001 range. In each year, the range was altered in a way that at one level can be seen as refining the differentiation among users and price levels but at another level can be viewed as forcing cyclists to depend on Shimano. Underlying the changes in the names and range of groupsets are much more significant changes in the designs of the components involved. Shimano undertakes continual refinement of its products to the extent that it was not possible in 1992 to order any component dating from before 1985. Furthermore, the trend toward component integration—most clearly evident in the STI (Shimano Total Integration) concept, which brings gear shifting, transmission, and braking together into one interlinked set of components—has meant that it is considered by observers to be increasingly difficult to mix Shimano components with those of another brand, or even with those of another Shimano groupset. In 1995, fears were expressed that a new transmission innovation, InterGlide chainwheels and sprockets, would prove incompatible with other companies' chains—another example of fears within the industry of what Shimano was capable of doing to its competitors. *Cycling Today* (September 1994: 13) saw InterGlide as a "very clever marketing strategy" that would capture the largest segment of the market—since it was being introduced initially among mid-range groupsets—and would force consumers into long-term dependence on Shimano for replacement components: "When your chain wears out, you will no longer be able to purchase a Sedis model to replace it; it won't be compatible. When your chainrings wear out, forget Pace, Specialized, Middleburn or TA chainrings; your choice is Shimano, Shimano or Shimano." The expected result of such changes was that Shimano's competition would be "crippled," with cycle shops and consumers left in a state of confusion about which components would go with which models of bicycle.

The example above highlights not only Shimano's innovatory skills and industrial ambitions but also the rest of the cycling world's *perceptions* of Shimano. Articles in the American cycling press, especially, portray an industry whose fear of the perceived power of Shimano had made few manufacturers willing to put their names to comments critical of the company (*Mountain Bike Action* 6, 1991, no. 8: 97). In the opinion of many of these anonymous cycle manufacturers, Shimano had become by the early 1990s an uncontrollable giant dictating every aspect of the industry's development (ibid.: 94–108; Espinoza 1992: 50–52). The company was then, and continues to be, seen by many in the industry as controlling the future of cycling—a view expressed in statements such as these:

> Shimano is the bicycle industry. . . . They dictate the state of performance. (anonymous product manager for an American manufacturer, quoted in *Mountain Bike Action* 6, 1991, no. 8: 97)

> [Shimano] could put anyone out of business if they wanted to. Shimano has so much money, it's scary how big they are. They could decide to lose money on a product to get rid of someone else and still easily come back and recoup their losses to control the market." (ibid.: 104)

In the United Kingdom, an individual who was working as a designer at Muddy Fox when that company went into receivership attributed the decision at Muddy Fox to go into liquidation in part to the introduction of Shimano's 1992 groupsets. The radical change from the previous year's range was seen as an attempt by Shimano to reduce overcapacity in the cycle industry by forcing less financially secure companies out of business by restricting the availability of the new groupsets, thus damaging the credibility of shops and manufacturers left with outdated stock.

The perceived attempt by Shimano to "shake out the dead wood of the industry and get rid of overcapacity" in a time of recession seems somewhat dubious, or at least exaggerated. The views put forward at Muddy Fox may well have been a means of shifting responsibility for the company's 1992 collapse away from its own financial problems (which, to judge from records held by Companies House in London, took precedence over industry politics in the decision to call in the receivers).[17] Though Shimano no doubt does pursue longer-term strategies than many other companies in the industry, it is in its own interests to promote the well-being rather than the decline of cycle manufacturers. Shimano may well "want to own the business" (in the words of a product manager quoted in Espinoza 1992: 50), but extracurricular activities such as sponsorship of a cycle museum in Sakai, promotion of sporting events, and support for cycle activism indicate that it wants the business to be successful.

The perceptions of Shimano reveal as much about the nature of the cycle industry as they do about Shimano. In many ways, the cycle industry today has many of the same traits that it had in the days when it was first beginning to adopt the processes and practices of factory production. Fears about Shimano reflect the same concerns that were common in that earlier period: concerns about a rapidly expanding and changing market, about a high rate of new entry and failure among producers, and about fickle consumers who might at any point switch to another manufacturer's products or even abandon cycling. Furthermore, in the past Shimano's attempts to simultaneously refine both its technology and its control of componentry supply led not only to the crippling of its major competitors (e.g., SunTour) but also to new opportunities for others to develop Shimano-compatible parts and accessories. Sociotechnical change cannot be directed by any one actor with complete certainty over the outcome. Social or political intentions built into technology can often be subverted (Akrich 1992).[18]

Built-In Obsolescence vs. Environmental and User Friendliness

Nevertheless, the cynical view of Shimano held by other industry players is shared by cyclists concerned with the repairability and replaceability of components. Shimano does a great deal to promote itself as a "caring," environmentally aware company. A video produced to celebrate its seventieth anniversary emphasized the importance of harmony among people, nature, and machines. The company was presented as creating "a bond between people and their machines" while striving to "improve the harmony between nature and humankind." The video closed with the slogan "Shimano—embracing our world, closer to nature, closer to people."

However, some of the strongest criticisms of Shimano focus on the difficulty of repairing and replacing products whose design incorporates planned obsolescence. STI is an archetypal black-boxed technology (Pinch and Bijker 1984). Its workings are beyond the understanding of most users. The combined brake-and-shift-lever units have been described as "so complicated that the manufacturers rightfully warn against taking them apart when they don't work properly: you'll have to replace the whole unit" (Van der Plas 1991: 154). This angers cyclists who like to repair components rather than be forced to replace a complete unit just because one part of it is broken. (Such complaints have been aired regularly in the letters pages of cycling magazines.) In this respect, Shimano components are seen as "particularly consumer hostile" (ibid.: 154–105), requiring upgrading and replacement rather than repair (*Cycling Today* October 1994: 16). The other side of this critique is that planned obsolescence generates a great deal of unnecessary waste. "What is so environmentally healthy about having to throw everything away as soon as it breaks?" asked one product manager (quoted in Espinoza 1992: 52).

Shimano's relations with its trade customers and its end users are fraught with contradictions. First, even its critics recognize the company's genuine commitment to developing high-quality components, yet this very same quality has been dangerous for the industry as a whole since the competition has been all but eliminated. Second, Shimano has been instrumental in getting more people onto bikes, yet at the same time it has made them helplessly dependent on its own equipment. Finally, Shimano has helped promote cycling as a means of integrating people within their environment, yet its products are not environmentally friendly. The most striking facet of the "Shimano paradox" is the way that it seems to inform every aspect of the company's activities.

Despite the negative aspects of what Shimano does, in the terminology of actor-network theory the company has been a highly successful network builder. To a degree this is not surprising, since Shimano has been extremely active and effective in building its network—particularly in setting itself up as an *obligatory passage point* (Callon 1986a, 1986b) for nearly every other element of the cycle market through strategies such as the development of user-friendly technology (for those who don't want to repair their own bikes) and integrated systems which are incompatible with other products. Other members of the network find that they cannot help but remain enrolled. Manufacturers are obliged to equip their products with Shimano components to avoid missing out on sales. Retailers must stock Shimano-equipped models for the same reason. Cyclists must have Shimano components to be like their friends, or because they "know" they're the best. Once one has a Shimano-equipped bike, each component must be upgraded with compatible Shimano equipment (or the bike must be stripped down and re-equipped with other components, an increasingly difficult task owing to the lack of availability and the poorer quality of other components other than the most expensive ones). The bikes themselves must be capable of taking Shimano fittings in order to be salable. Magazines may criticize Shimano, but they are dependent on the firm for advertising revenue; in any case, they cannot help but publicize the activities of an omnipresent company.

What is remarkable about this network is that it is by no means Shimano alone that does the work of maintaining it. Consumer demand, the generally high performance of the components, favorable test reports in magazines, and the desirability of the Shimano label seem sufficient to keep resistant manufacturers tied into the network despite their ambivalence (Singleton and Michael 1993). Shimano is then at liberty to produce innovations that further tie consumers and manufacturers into the network. The complexity of the relations among manufacturers, retailers, and consumers within Shimano's network illustrates the obduracy of sociotechnology; while this does not make Shimano's dominance irreversible, the company is a central element in the sociotechnical ensemble of the bicycle, and breaking up its network will be very difficult (Law and Bijker 1992; Bijker 1995).

Nevertheless, there is resistance in this network, manifested in the attempts by many manufacturers to equip one or two models each year with other companies' components to try to break Shimano's stranglehold. This strategy wasn't helped by SunTour's loss of a UK distributor in 1993, but it was boosted shortly afterward by the market growth of SRAM's Gripshift gear changers in the middle and lower price ranges. The purchase of Raleigh's former gear department (Sturmey-Archer) by the Taiwanese firm SunRace may give that firm credibility as a competitor for Shimano.

Shimano has not taken such challenges lying down. It responded to the inroads made into its market by SRAM with legal action in the European Union. However, the willingness of Raleigh and some other companies to use Gripshift on a significant proportion of their products, and the willingness of consumers to buy these bicycles rather than Shimano-equipped models, indicates that the Shimano stranglehold may yet loosen, at least in the area of gear shifting. SRAM's strength may be enhanced by its 1997 purchase of the German component manufacturer Sachs.

Shimano's role within the bicycle industry is, then, highly paradoxical. Its activities involve a number of interlinked contradictions—between the innovation and change that characterize its business strategies and the longevity, consistency, and reliability of products that some users require; between the environmental discourses used by all actors in the world of cycling and the built-in obsolescence that is a feature of Shimano products; between the company's success in promoting cycling to new consumer markets and its relations with the rest of the industry (which feels unwillingly locked into Shimano's product network).

These contradictions contrast sharply with the pairings of opposing values that shaped the mid-twentieth-century bicycle industry—which was centered on mass production and consumption, modernity, and the city—and the industry's relations of production. In the sociotechnical frame of the globally flexible bicycle, these issues are largely uncontested, as a result of the transformations of cycling and the bicycle industry that have taken place since the 1970s. Production methods have been thrown open through the widespread adoption of the discourses and the methods of flexibility, and through the globalization that Shimano's industrial strategies exemplify. At the same time, the importance of production within the frame has been tempered by the increased role of consumers in shaping products and their meanings. Struggles over the relations of production have been superseded by uncertainty about the industry's future, a situation which is again closely linked to the global pursuit of cheap labor. The greatest consistency between the mass bicycle and the globally flexible bicycle lies, then, in the cultural meanings they hold. While these have changed from one sociotechnical frame to the next, it is not difficult to identify similarities in their shared concern with modernity, nature, and the city.

Culture and Change in the Sociotechnical Frame of the Globally Flexible Bicycle

I have suggested a number of times now that a primary difference between Bijker's model of sociotechnical change and the one I have been developing in this book is the role of culture in shaping change. The sociotechnical frame of the globally flexible bicycle is full of examples of this, since cycling cultures have been crucial to the way new configurations of the bicycle have emerged alongside, capitalized on, and further stimulated the changes in production and organization that have characterized the cycle industry since the 1970s. This is especially true of the cultures of mountain biking, which have helped to construct a highly diverse and innovation-hungry consuming public that matches the newly acquired capacity of the industry to meet changing demands more flexibly, quickly, and "efficiently." These demands range from the rough off-road sporting heritage of the original mountain bikers through more conventional sporting and leisure riding to family day trips and urban commuting. These different activities generate a myriad of cycling cultures, including one centered on sport and one devoted to off-road leisure. It also generates vociferous cycling advocacy focused on transportation policy, urban planning, and environmentalism.

The Cultural Contradictions of Mountain Biking

The concepts and debates concerned with modernity and postmodernity that have formed the backdrop of my discussion have been central to the shaping of cycle production and cycling culture. To begin with, the bicycle was itself a factor in the shaping of modernity, playing a major transitional role in the simultaneous development of mass production and mass individual transportation (Hounshell 1980, 1984; Norcliffe 1997; McGurn 1999). The freedom of movement provided by cycling opened people's minds to the opportunities for achieving greater speed with less effort that soon appeared with the automobile. At the same time, refinements in manufacturing techniques developed by Pope, the Western Wheel Works, and others paved the way for a mass-production industry to support the new desire for individual mobility. As Harvey mentions (1989: 264), bicycles were an important element in the changing spatialization and temporality that characterized modernity in the late nineteenth century.

The relation between cycling and modernity continued in the twentieth century, as is evident from Sir Harold Bowden's writings (discussed in chapter 3 above). Discourses of cycling and of cycle manufacture display an ambivalence toward modernity, technology, industry, and urban life (Patton 1993)—for example, in espousing the latest technology as a means of escaping the same civilization that created it. This dichotomy is not just characteristic of the interwar years; it is also found in discourses associated with mountain biking, going back to the sport's origins in California in the 1970s.

Mountain biking juxtaposes the pursuit of frontiers and of supposedly authentic experiences of the wilderness with highly sophisticated new technology. Mountain bike activists and campaigners are highly conscious of this contrast, which is prominent in debates with hikers, equestrians, and other users of wilderness areas. One speaker told a 1992 conference on mountain biking and the environment in the Lake District how "the sight of a single deep wheel track during a day's walking on the Howgill Fells appalled him and filled him with a deep sense of hatred." For this speaker, "our respect and love of the wilderness comes from getting as far away as possible from modern society and especially from all machines" (Colin Mortlake, quoted in *Westmoreland Gazette*, February 21, 1992: 11; see also Wyatt 1992). Such a view defines bicycles—especially mountain bikes—as "machines," in contrast to less objectionable "tools," such as climbing equipment, canoes, and skis (ibid.). For others, the qualitative difference between the rubber of a bicycle tire and the rubber of a walking boot is far less significant than that between mountain bikes and motorized transportation (letter in *Westmoreland Gazette*, February 28, 1992: 13).

This question of the meaning of technology and its relationship to the environment lies at the core of mountain biking, both for mountain bikers and for those who find their presence intrusive. Mountain bikers counter complaints about environmental damage by arguing that wheel tracks cause no more damage, and perhaps less, than walking boots and horses' hooves. A 1987 report by Santa Barbara Ranger District regarded the erosion caused by mountain bikes as acceptable in comparison to erosion caused by other activities (Gibbs 1992: 42). As in the Lake District, objections to mountain bikes by other groups was found to be more aesthetic than environmental, concerned primarily with the perception of cyclists bringing "city values" into the "wilderness" (McNaghton and Urry 1998). Where damage has occurred, in both settings, it is to do with the use of unauthorized routes and with the number of users rather than with the nature of the recreation itself. In Britain, attempts to counter the overuse of trails have resulted in the establishment of a Mountain Bike Code of Conduct and in voluntary bans on mountain biking on certain routes during peak times. Cycling magazines also feature regular articles on environmental issues with the aim of drawing the same links between cycling and "green consciousness" that are raised in the rhetoric of Shimano.

Mountain bikers situate themselves, then, within the often-contradictory discourses of the new environmental social movements that have emerged since the 1960s—movements concerned with nature and "wilderness" on the one hand and with urban land use and transportation on the other (Eyerman and Jamison 1991; Wall 1999). Like these various strands of environmentalism, mountain biking (and recreational cycling more generally) constitutes a rejection of modernity that is at the same time dependent on modernity for the very basis of its existence. This contradiction, with its "juxtaposition of diverse and seemingly incongruous elements" (Harvey 1989: 338), situates mountain biking within the shifts from modernity to postmodernity, from modernism to postmodernism, and from modernization to postmodernization.

Of particular relevance is how the growth of mountain biking among baby boomers in the 1980s drew on "willful nostalgia," a major element of postmodern culture (Jameson 1991; Robertson 1990). Early mountain bikes, going back to the "clunkers" of Ritchey, Kelly, and others, can be understood in relation to nostalgia for the bicycles of youth. Especially pertinent here is Fredrik Jameson's notion of "pastiche." Jameson uses this notion to define the nostalgia films of the 1980s, which he describes as a "desperate attempt to appropriate a missing past," most notably "the henceforth mesmerizing lost reality of the Eisenhower era." He argues that "for Americans at least, the 1950s remain the privileged lost object of desire—not merely the stability and prosperity of a pax Americana but also the first naive innocence of the countercultural impulses of early rock and roll and youth gangs" (Jameson 1991: 19). This is clearly resonant with the early mountain biking culture,[19] as are Jameson's comments on "the incompatibility of a postmodernist 'nostalgia' art language with genuine historicity" (ibid.). Jameson argues that contradictions "propel [the nostalgia film] into complex and interesting new formal inventiveness; it being understood that the nostalgia film was never a matter of some old-fashioned 'representation' of historical content, but instead approached the 'past' through stylistic connotation, conveying 'pastness' by the glossy qualities of the image, and '1930s-ness' or '1950s-ness' by the attributes of fashion" (ibid.).

Like nostalgic films, clunkers were 1970s remakes of a 1950s artifact, and the quotations earlier in this chapter show that this is clearly how they were read culturally by many early users. As the technical aspects of mountain bikes removed them further and further physically from their 1950s reference point (the Schwinn Excelsior), the symbolic level of their "1950s-ness" become more prominent, at least until the American and British mountain bike booms diffused the coherence of mountain biking culture. Mountain bikes can be regarded, then, as artifacts of postmodern culture, while mountain biker identities must be seen as having been constructed through the same struggle between modernity and postmodernity as the bikes themselves (Rosen 1993).

Production and Consumption: Sociotechnical Change from the Mass Bicycle to the Globally Flexible Bicycle

The cultural situatedness described above informs the production as well as the consumption of bicycles. The changing practices of management and manufacturing that have underlain the bicycle industry's shifts in sociotechnical frame stem from wider cultural changes within management thinking that go beyond any particular industry. Harvey (1989: 28) draws an explicit link between modernity and mass production, seeing the publication of Taylor's *Principles of Scientific Management* and the launch of Ford's moving assembly line as parts of the growth of modernist culture in the early twentieth century. Likewise, Harvey regards the shift to what he terms "flexible accumulation" as going hand in hand with the rise of postmodernism. The latter two are characterized by acceptance—in contrast with modernism's rejection—of fragmentation, ephemerality, uncertainty, and multivocality (ibid.: 44). Postmodern culture celebrates these features in the juxtaposition of different styles in a single building, in the superimposition of one work of art over another, and in the attachment of modern components to twenty-year-old bicycle frames. Similarly, flexible manufacture has introduced methods such as just-in-time inventory control and CAD/CAM as means of capitalizing (literally) on fragmented and ever-changing markets, rather than seeing these market changes as a threat to existing production scales. The shift to globalized and more flexible manufacturing approaches can be seen as a means of adapting positively to difficult "new times." (Compare Hall and Jacques 1989.)

The paradox of the British cycle industry's relationship to postmodernity is that the factors that simultaneously triggered both postmodernity and Raleigh's economic decline in the 1970s have at the same time boosted cycling at a cultural level, thus allowing emerging industries overseas to capitalize on Western misfortune. Harvey (1989: 328) famously dates the start of postmodernity to 1973, the year of the first of the two oil crises of the 1970s. The second of these crises, in 1979, was a major factor in the loss of world markets that precipitated Raleigh's decline, yet the shortages of gasoline that these crises generated in the West helped promote cycling as a means of transportation. The ecological damage caused by industrial production has similarly helped promote cycling as an environment-friendly activity, and the cycle industry has been quick to claim environmental credentials for innovations in their factories that minimize harm to the environment and to the health of workers. In fact, many of these innovations are required by European Community regulations anyway, and in the past cycle production was by no means environmentally benign. Raleigh committee minutes as early as 1919 report complaints about factory fumes from local gardeners.[20] In the 1970s a report condemned the Raleigh factory's pollution of the surrounding area (Nottingham Workshop 1978), and Raleigh continued to be accused of causing pollution in the 1990s, although the company disputed this.[21] Nevertheless, personnel at all levels of the cycle industry—just like their customers—express strong support for environmental objectives and proclaim the environmental benefits of cycling relative to other modes of transportation.

Bicycle production and bicycle consumption are, then, both informed by a range of cultural influences. Some of these are specific to bicycles and cycling, focused on particular ways of using the technology—e.g., for transportation, for leisure, for sport, for pushing the boundaries of experience in the wilderness. Others are concerned with broader matters, such as the benefits of certain approaches to production and management and the relation of technology to modernity, postmodernity, and the environment. These issues have cohered in particular with the globally flexible production of mountain bikes, where the mutual influences of this sociotechnical frame and associated cultures are especially clear. The knowledges, practices, methods, techniques, and relations of social actors, artifacts, and production technologies that make up this particular sociotechnical frame of the bicycle are closely interlinked with cultures that express the "ways" (Traweek 1992) of mountain bikers, cycle couriers, commuters, leisure cyclists, and cycling activists, alongside the advocates of new theories of management, organization, and production.

The thorough integration of culture within the frame of the globally flexible bicycle throws light on how this frame emerged out of the earlier frame of the mass bicycle. Again, this shift can be accounted for by combining Bijker's two mechanisms of change: (1) the encounter of the established frame with alternative approaches and (2) the low inclusion of significant actors within the established frame who stimulate the development of a new one which borrows features from both the established and alternative frames. Both of these processes can be seen in the shift from the mass bicycle to the globally flexible bicycle, spurred on and mediated by the cultural changes I have been discussing. (See figure 6.7.) During the 1970s and the 1980s, the sociotechnical frame of the mass bicycle was confronted with the emergence and rapid growth of a new frame originating in the Far East, alongside the changing management

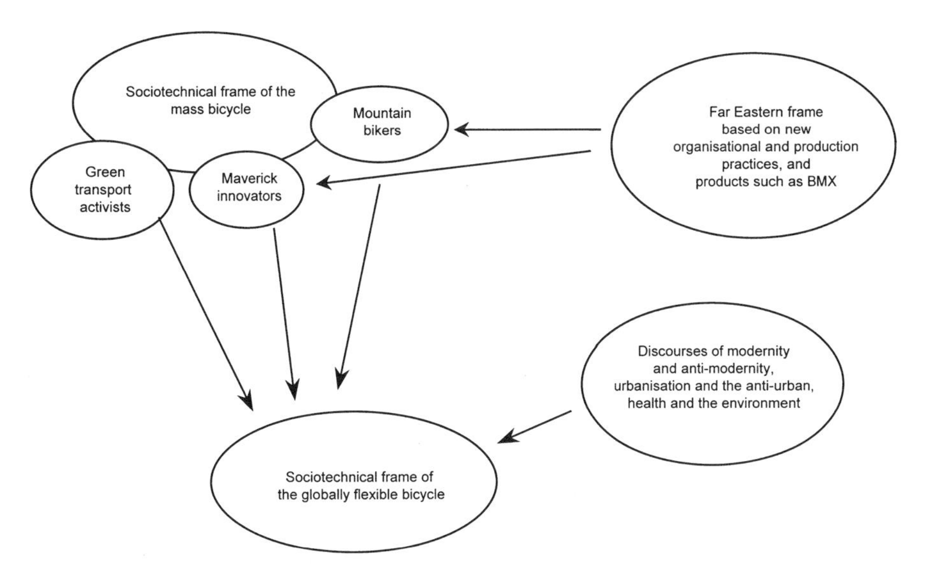

Figure 6.7
The shift from the mass bicycle to the globally flexible bicycle.

cultures that were then promoting flexibility. At the same time, a number of actors with low inclusion in the established frame of mass production gained prominence in the cycling world and in the cycling culture through the growing popularity of environmental and health concerns. These included early mountain bikers, green activists, radical innovators, and athletes.[22] The encounters of these marginal groups with the new frame helped to stimulate the shift to flexible manufacture.

The point just made can be illustrated with the example of Far Eastern production methods, which have now permeated the British (indeed the global) bicycle industry. Central to the Taiwanese approach to cycle production is the use of flexible technologies, such as TIG welding of frame tubes. This technology provides a good example of what happens when a marginal social group encounters an alternative sociotechnical frame. TIG welding was new to Western cycle production and was not easily compatible with the traditional style of frame building, which involved brazing bicycle tubes together inside sleeves (known as "lugs") that enclose the tubes' ends. With TIG welding, lugs are not used. Instead, the tubes themselves are welded directly together in a blanket of inert gas, using a tung-

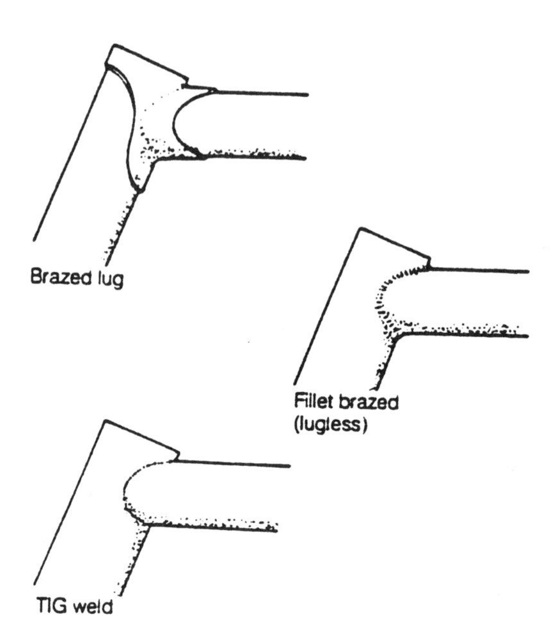

Figure 6.8
Three ways to weld a bicycle frame. Reproduced with permission from Van der Plas 1991.

sten welding element.[23] This method appears to have been introduced into cycle manufacture from the aerospace industry by Californian builders of BMX bikes in the early 1980s. In fact, the development of BMX parallels that of mountain bikes in many ways. As with the early development of mountain bikes, BMX was a grassroots innovation and activity that quickly spread into mainstream cycle production and cycling culture. The differences were (1) that BMX was always clearly a part of youth culture rather than the adult markets at which mountain bikes were initially aimed and (2) that after a boom in 1983–84 BMX sales declined quickly, leaving BMX a much smaller-scale "cult" activity. In contrast, mountain bikes have expanded far beyond their original target market. Nevertheless, much of the technological innovation of the early mountain bike years was derived from BMX, and indeed "Marin county conversion" of clunker bikes could not have happened without the appearance in the late 1970s of new BMX components. BMX may even have been the initial source of the geographic dispersal of cycle production into the Far East.[24]

TIG welding is conducive to flexible production at a number of levels. It is quicker and requires far less skill than brazing, so TIG-welded bikes are cheaper to produce. It also allows a more flexible approach to design than brazing. Lugs are produced in batches of thousands, at a fixed angle. This makes them well suited to the huge runs of standardized designs in mass production, but not to the niche marketing of flexible manufacture (where an individual model might be produced only in hundreds or even only in tens). TIG welding has generally been the favored joining method for mountain bike production, although the earliest mountain bikes were lugged. The flexibility of frame angles that TIG welding facilitates allowed for a gradual shift in mountain bike design over the course of the 1980s, aiding the spread of this new bicycle design out of its initial specialist niche (see Rosen 1993). Early mountain bikes followed the frame geometry of the Schwinn bikes on which they were modeled—the angles of the head and seat tubes were both "slack," laid back in an "easy rider" position (with head-tube angles typically as low as 68° and with seat-tube angles of 70°; see Kelly and Crane 1990: 90)—a position that was ideal for the fast downhill riding of the "repack run." By about 1989, both angles had been gradually brought closer to those of conventional road bikes—about 73° for the seat and 71° for the head (Bogdanovich 1989). Mountain bikes thus became suitable not just for downhill racing but also for a variety of uses on and off the road.

Aside from the utility of different frame angles, frame geometry serves a rhetorical purpose in manufacturers' catalogs in indicating the style, the attitude, and even the presumed gender of the rider. This flexibility of design was made possible by the use of TIG welding, and hence by the integration within established production centers of new Far East-derived production methods. TIG welding has thus served both technologically and culturally in facilitating a proliferation of different cycling uses and cultures through the proliferation of design choices.

Underlying the shift just described has been a revision of how organizations in the cycle industry see themselves working. The heterogeneity of actors and practices involved in globalized flexible cycle manufacture shows that this frame is by no means experienced in the same way by all participants. Flexible production covers a wide range of approaches and methods: the huge output of various models for a diverse range of clients that is typical of Taiwanese factories; the more coherent, though still flexible, output of Raleigh, which until recently was producing 500 at a time of each model in one factory but much smaller batches in the other; the comparatively tiny batches produced by smaller manufacturers. Even this set of characterizations masks differentiation.

Raleigh's response to the growing influence of the Far Eastern sociotechnical frame has gone through a number of stages. As imports rose and exports dropped in the 1970s and the early 1980s, Raleigh began to find itself becoming marginal to the direction in which the industry was developing. The failure of TI's initial response to this decline (trying to bolster the mass frame with more investment) was eventually replaced with a more radical solution—a transformation of the organization—which then brought British cycle production out of its decline and into a new frame that drew much inspiration from the Far East. The sociotechnical frame of the mass bicycle was effectively abandoned, although many of its central features remained in the new frame.

The shift to globally flexibilized manufacture thus came about to a large extent through a combination of the increasing marginalization on an international scale of the British cycle industry establishment and the simultaneous rise to prominence of several new sets of interests and cultures. This follows the pattern of change in the sociotechnology of the bicycle as the factory bicycle was replaced by the emergence of mass production and mass consumption during the 1930s, the 1940s, and the 1950s. Might this pattern be repeated as interests and cultures concerned with the bicycle again begin to shift, this time toward concerns about sustainability? What might form the basis of a new sociotechnical frame of the bicycle?

7

Up the Velorution[1]: Toward a Sociotechnical Frame of the Sustainable Bicycle

Dynamics and Politics of Sociotechnical Change

In the preceding chapters, I have traced historically the changing relations of technology, society, and culture in relation to the bicycle, focusing on the sociotechnical frames of the factory bicycle, the mass bicycle, and the globally flexible bicycle. This process has also featured a number of significant transitional periods *between* sociotechnical frames when instability in the configuration of the sociotechnology of the bicycle has led to the decline of one sociotechnical frame and the eventual formation of another. Just as studying the out of the ordinary often allows a better understanding of the everyday in disciplines such as psychology (Freud 1901) and anthropology (Turner 1957), shifts from one sociotechnical frame to another clarify the features of these frames during periods of stability.

Bijker (1995: 14) expects a good theory of sociotechnical change to be able to explain the constancy and continuity that persist alongside flux and change. The transitions between frames often throw light on how certain features remain a part of a *sociotechnical ensemble* despite the changes that are taking place. One prominent continuity in the story of the bicycle has been the ambivalence toward modernity that has remained a consistent element of bicycle sociotechnology from mass production to global flexibilization, even though the form of this ambivalence and the ways in which it is realized have differed from one frame to the next. Fleeing modernity for the wilderness by bicycle is now regarded as a more visibly technological activity than it was previously, and there seems to be less of a concern among wilderness cyclists of the 1990s than among those of the pre-World War II years to hark back to a pre-modern way of life in doing so. The revised SCOT approach I have been taking in this book is able to locate such continuities and differences in the changing relationship of cycling cultures to wider cultural values and to bicycle technology and the social world of the bicycle.

Paying attention to the continuity of sociotechnical change opens the way to understanding stability and obduracy. The stabilization of an artifact leads to the subsequent stabilization of a whole network of relations of technology, society, and culture. The artifact's relation to other technologies in the same sociotechnical frame, the establishment of technical and social infrastructures to support it, and the embedding of these infrastructures within cultural practices all serve to make that frame obdurate. This is most clearly visible in the case of the automobile, whose obduracy as a technology has been sustained by a parallel stabilization of highway infrastructure and urban design; by social and cultural practices in relation to the home, work, and leisure; and by the cultural embedding of values associated with the apparent freedoms provided by automobile ownership. The obduracy of the automobile lies in the complexity of this *ensemble* of technologies, practices, and values. To disembed the automobile from Western culture would entail disembedding each of these different elements from the overall *ensemble*—an extremely difficult task.

Examples in the sociology and the history of technology show us, though, that difficult or unlikely tasks are not impossible, and that what appear to be the most solid of sociotechnologies can collapse. (See, for example, the work of Callon.) In the case of the bicycle, what at one time appeared to be an irreversible trajectory of mass cycle production for an ever-expanding market turned out to be a relatively short-lived situation. However much personal and organization investment—of skill and expertise, of money, of time, of the identities of those involved—has gone into establishing a sociotechnical frame, unanticipated factors can lead to its demise. Consequently, the obduracy of sociotechnology is always temporary; instability seems an inevitable part of the process, playing a considerable role in the shaping of an artifact.

The dynamic nature of the sociotechnology of the bicycle has transformed it first from a craft product aimed at elite leisure users toward a combined system of mass production and consumption, mass transportation, and an associated leisure culture characterized by an ambivalence toward modernity and the city, then to a new configuration where more fragmented and somewhat contradictory leisure cultures have been supported by a fragmented, highly innovative, globally dispersed industry. The less stable periods in which these transformations have taken place are valuable in illuminating the processes of change, highlighting the less robust elements of the frame that lead participants to search for alternatives. They also make visible the wider cultural changes that often underlie sociotechnical change, and which subsequently inform the shaping of a new frame. In the case of the mass bicycle, this was most obvious in relation to the way that the standardization of products which was necessitated by manufacturers' production decisions became increasingly out of step with a cycling culture that sought variability of goods and product innovations to satisfy its changing consumption demands. This cycling culture must be set within a broader cultural context of late-twentieth-century work, leisure, and consumption patterns—in short, increased automobile ownership was a major factor in the way that cycling ceased being a significant means of commuting, and leisure cycling came to account for the bulk of cycling activity.

While the globally flexible bicycle represents a notable advance for consumers in terms of their potential input into design decisions, presenting instability as an integral feature of sociotechnology also offers the hope of further change to those who would question how equitably the benefits of global flexibilization are distributed, both within the bicycle industry but also (for the economy and politics) more broadly. Within this new sociotechnical frame, bicycles are both a material and a cultural product of complex global networks of producers, assemblers, distributors, and marketers—a product which is then consumed by markets which are equally complex and fragmented, and constituted in large part through the culture of postmodernity. The ambivalence of various commentators toward this global flexibilization of production, consumption, economy, and culture is grounded in concerns that would have been familiar to Marx—concerns about who benefits and who suffers from it, about *how* these benefits and disbenefits are experienced, and about the role of technology in facilitating the changes that have brought it about. While Bookchin (1974) argues that automation can benefit local communities and free them from unnecessary toil, for others (e.g.. Hobson 1992; Garrahan and Stewart 1993) the interests of capital have acted to ensure their own benefits to the detriment of workers long before any community benefits might have become apparent.

Bookchin's (1974: 86) concern to identify "liberatory technology" focuses on three questions:

> What is the liberatory potential of modern technology, both materially and spiritually? What tendencies, if any, are reshaping the machine for use in an organic, human-oriented society? And finally, how can the new technology and resources be used in an ecological manner—that is, to promote the balance of nature, the full development of natural regions, and the creation of organic, humanistic communities?

To transform the sociotechnology of the bicycle into something that meets these concerns would demand changes in how various interests are configured in the social world of the bicycle. Workers in the industry might strive for more equitable control of production, alongside marginalized designers and users of bicycles (e.g., women, the disabled, and those concerned with utility rather than leisure designs) The requirements of such groups are not so easily met through innovations intended to increase productivity and to simplify design for greater profitability. The kind of access to decisions about design and industrial strategy that such groups aspire to are also the focus of cycling cultures concerned with health, the environment, and transportation. These issues have played an important discursive role in stimulating cycling under global flexibility. They could have a more substantial impact on how the sociotechnology of the bicycle develops beyond global flexibility if allied to marginalized interests within the production and consumption of bicycles.

The Bicycle and Regulation

So far in this book I have paid little attention to regulation or government policy. My neglect of this topic up to this point is, I believe, defensible in that in the United Kingdom there had barely been any policy relating to cycling apart from safety standards and the prohibition of bicycles from certain roads until the 1980s. In the United Kingdom, cycling has been regulated to a certain extent since the late nineteenth century, both by the state and by cycling bodies such as the Union Cycliste Internationale (UCI) and its UK member organizations. The 1888 Local Government Act was especially important in defining bicycles as carriages under the terms of the 1878 Highways Act. (See Rosen 1995a: 109–110.) The Highways Act had already resulted in many cyclists' being charged for "furious driving" (Woodforde 1970; Alderson 1972). The 1888 act required cyclists to carry a lamp between sunset and dawn and to sound a bell or a whistle when approaching other road users (*Times*, August 22, 1888).[2] Such legislation drove road racing underground and led to the emergence of time trialing as a means of overcoming restrictions on massed-start races (Alderson 1972; Watson and Gray 1978). However, until recently there was very little in the way of UK government policy toward cycling, except by default—that is, as a by-product of transportation policies that have increasingly promoted the use of private automobiles for commuting, for leisure trips, for shopping, and so on, and have thus marginalized the use of the bicycle as a credible means of transportation. This relegation of cycling to the margins—as a children's toy or as a high-tech leisure apparatus for off-road use—has coincided with, and thus played a role in shaping, the emergence of the globally flexible frame of the bicycle. However, an equally important component of that frame has been the strong concern with "wilderness," with the environment, and with the congestion and health impacts of transportation. Since the late 1980s these have been increasingly important focuses of government policy too, and as a result cycling has increasingly been seen as a solution to transportation problems. While environmental concerns have not yet been translated in the United Kingdom into a significant rise in the use of bicycles for transportation, might the policy shifts of recent years—in conjunction with growing environmental concern across British society and culture—trigger a transition to a new sociotechnical frame of the bicycle in which such issues are central rather than incidental?

Transportation, Sustainability, and Mobility

Sustainability in the International Policy Arena

The concept of sustainability has become increasingly important in international policy debates—especially since the publication of the Brundtland Report (World Commission on Environment and Development 1987), which popularized the concept of "sustainable development" ("development that meets the needs of the present without compromising the ability of future generations to meet their own needs"—ibid.: 43). There has been much debate about the soundness of sustainability as a concept; about the Brundtland Report's framing of sustainability in terms of the equally problematical notion of "needs"; about the problems that arise in trying to define "sustainability," "development," "sustainable development," "the environment," and so on; and about the likelihood that competing interpretations and competing interests among the parties to international policy discussions, such as governments, non-government organizations (NGOs), industry, and banks can be reconciled There has also been a great deal of tension between the perspectives of "Northern" and "Southern" NGOs and governments, which came to light particularly at the 1992 "Earth Summit" in Rio de Janeiro.[3] Nevertheless, "sustainability" has acquired widespread recognition as a means of expressing a way of approaching development that aims to be environmentally, socially, and culturally benign and is sensitive to a wide range of needs and interests, both contemporary and longer-term.

Sustainability is a concept that can be usefully applied to matters other than development. Most notably for this book, a significant strand in the sustainability agenda that developed through the late 1980s and the 1990s has been transportation. Agenda 21, the international sustainability document adopted at the Earth Summit, identifies transportation as an issue to be addressed in the move toward sustainable development,[4] highlighting resource depletion in the production and consumption of cars, emissions in their use, patterns of urban design that encourage traveling, and the impact of transportation on urban populations.

The various aspects of how transportation interacts with the environment that are identified in Agenda 21 are already important concerns in the developed world, but they are potentially just as crucial in the developing world as the demand there for transportation and related infrastructures expands. Consequently, the document calls for a combination of measures to address them in both the North and the South—technological solutions in the form of fuel efficiency and emissions control, policy measures that deal with land-use and transportation planning, fiscal means to encourage the use of more sustainable modes of transportation, the promotion of public and non-motorized forms of transportation, and so on. Although the effectiveness of the Rio Summit was being questioned even as it was taking place, its transportation-related proposals were already becoming common currency among transportation professionals.

Sustainable Mobility in the United Kingdom

The beginnings of a sustainable transportation policy in the United Kingdom are evident in This Common Inheritance, a 1990 government white paper on sustainable development (Department of the Environment 1990). The transportation section in this white paper was framed (as was the whole document) in terms of market forces' meeting the needs of sovereign, rational travelers in response to increased demands for environmentally less damaging forms of transportation. This was to be underpinned by a small degree of regulation to encourage a "modal shift" from the private automobile to other modes of transportation.

Aside from the questionable assumption that the market can be effective as a driving force for change, this approach has implicit problems in treating transportation individualistically (Lutzenhiser and Shove n.d.). In order to assess and then change travel habits, it is crucial to pay attention to trends and patterns across a town, a region, or indeed a country rather than just to individuals. Nevertheless, This Common Inheritance did include the seeds for more radical developments—for example, in referring to discussions then taking place about the relationship between transportation and land-use planning. These discussions came to fruition later in the 1990s, most notably with the revision of the Department of the Environment's "planning policy guidance" notes for transportation, town centers, and retail development (Departments of Environment and Transport 1994; Department of the Environment 1996). The new guidance notes were expressly concerned with reducing automobile traffic and encouraging the use of other modes of transportation, and they highlighted the traffic caused by new developments such as out-of-town retailing, housing, and leisure facilities. Land-use planning, in particular, was identified as a mechanism by which the traffic impacts of such developments could be regulated. This marked a shift away from the laissez-faire development policies of the 1980s.

These new policy guidelines appeared within a broader policy context in which "sustainable transportation" was becoming increasingly significant, signaled by a succession of reports and government publications that favored a more integrated approach to land use and transportation planning, a modal shift away from the automobile, and a reduction of transport's impact on the natural, social, and urban environments.

The 1994 Report of the Standing Advisory Committee on Trunk Road Assessment marked a major breakthrough in UK transportation policy, since it constituted the first official recognition that the building of new roads generates new traffic rather than simply catering to existing traffic levels (SACTRA 1994). The Royal Commission on Environmental Pollution's 1995 report on Transport and the Environment made clear its commitment to a sustainable transportation agenda and highlighted the unsustainability of current transportation patterns. It identified automobile usage as a major cause of environmental pollution and of poor quality of life, especially in urban areas, and it set out a comprehensive program of far-reaching policy recommendations to address these problems. Importantly, it followed the lead of Agenda 21 in taking a comprehensive approach to transportation, addressing both the production and the use of automobiles and other vehicles that have high resource consumption in their manufacture as well in their fuel use.

In 1995, in this climate of critical debate about transportation, Transportation Minister Brian Mawhinney set a "great transport debate" in motion with a series of speeches on transportation issues and with a national consultation that resulted in the publication of a government policy statement (Department of Transport 1996a) boldly entitled Transport: The Way Forward. This document endorsed the arguments of the Royal Commission without accepting all its recommendations; nevertheless, it indicated that sustainability was firmly on the transportation agenda by this point and that the prominence of market forces as a mechanism for change in the 1990 white paper had become strongly tempered by the need for government action across a wide range of levels and areas. The diminishing role attributed to the market was consolidated with the election of a Labour government in 1997, which published a transportation white paper (DETR 1998) in the following year—the first in more than twenty years—spearheaded by the new Deputy Prime Minister, John Prescott. This document built on previous policy developments in setting out a transportation policy that was intended to be integrated between different modes of transportation, with the environment, with land-use planning, and with other aspects of government policy (ibid.). While it is important to note its continuity with what had been happening for ten years, the white paper was broadly welcomed as an important (if still not fully developed) breakthrough in transportation policy by environmentalists and transportation activists. It is worth noting, however, that even when canceling planned road schemes the Labour government mimicked its predecessor in citing costs rather than policy as the reason, and that it was quick to back down on several commitments in the face of opposition. The lack of a firm shift toward a sustainable transportation policy was due partly to differences between the new Department of Environment, Transport and the Regions and other government offices (most notably the Treasury) and partly to a concern not to alienate voters by appearing to be "anti-car" (Rosen 2002a).

Road Developments and Road Protests

Others were perfectly happy to proclaim themselves "anti-car" during the 1990s. Despite the rapidly changing policy context, people in the United Kingdom were faced throughout the decade with continued road building and development, which appeared to contravene the latest guidelines and yet were permitted by the planning authorities. This was caused by a lag between policy and the planning process, in that a development that had received planning permission in the 1980s or the early 1990s could remain unbuilt for several years. Once development began, it would be taking place against a background of newer policies that no longer supported that particular kind of development. Consequently, in the 1990s local communities often found themselves contesting the construction of developments that contravened current policy thinking but were legally allowed since permission had been granted before the appearance of new guidance notes. One effect of this has been that the UK transportation agenda of the 1990s was dominated in the public eye not so much by the policy makers' gradual shift toward sustainability as by the high profile of protests against large-scale development, especially road building. The Conservative government unquestionably was seen—and saw itself—as the party of the automobile and road lobbies. Its comprehensive road-building program of the early 1990s was seen by many as a continuation of ten years' partisan support and subsidization of the motoring lobby (Media Natura 1990). This support was achieved in large part by running down public transportation—the Conservatives reversed the Greater London Council's early-1980s policy of cheap bus and underground fares, they failed to invest in the rail network, and they deregulated buses outside London so that the most popular routes became subject to competition while less popular "public service" routes were left to deteriorate.

The announcement in 1989 of a major new road-building program (Department of Transport 1989) merely served to confirm Prime Minister Margaret Thatcher's championing of "the great car economy" and her disregard of the people who were dependent on other modes of transportation and of the environment, which was then becoming widely recognized as under threat from automobile traffic and road building. Consequently, as the environmental and social impacts of the automobile became more widely understood and recognized through the 1990s, the road program was getting into full swing. The result was a clash at two levels:

- in a political and ideological sense, between existing and emerging policy objectives—i.e., between "the great car economy" and notions of sustainable transportation
- at a physical level, between road builders and developers turning out-of-date policies into material reality—by beginning to build new developments that had been approved some years earlier—and transportation activists who were not prepared to see these discredited policies implemented.

Beginning around 1993, highly publicized clashes took place over a number of construction sites, including the M3 link at Twyford Down in Hampshire, the M11 link at Wanstead, Solsbury Hill near Bath, the Newbury Bypass, and the A30 at Fairmile.[5] Without endorsing all the values of the protesters, the widespread public support for this steady stream of road protests showed at the very least a strong popular opposition to the building of roads in environmentally sensitive locations.

The 1990s also saw the emergence of tactics that questioned more generally the logic of an automobile-based transportation policy. A group calling itself Reclaim The Streets (RTS) would close off roads for an afternoon in order to hold illegal street parties that simultaneously highlighted the introduction of increasingly repressive laws against public protest and the negative impact of automobile traffic on street life. In a related strategy, "critical mass" bicycle rides around congested city centers demonstrated the advantages of cycling over dependence on the automobile. Another approach was to "develop" land in ways deemed more sustainable—notable here was the establishment by the group The Land Is Ours (TLIO) of an "eco-village" on derelict land in Wandsworth, South London, owned by Guinness, which had twice been refused planning permission for a retail development. These tactics often came together in mixed protests.[6]

Although the general public appears to have broadly supported transportation-related direct action (albeit with some qualifications), official responses have been less welcoming, and often violent. This has frequently led both the protesters and the authorities (or their agents) to become more firmly entrenched in their positions, one based on strictly legal arguments ("the development has planning permission, and is therefore legitimate") and the other on ethical or environmental grounds ("the development is harmful to the environment and thus isn't legitimate")—something that makes conflict increasingly difficult to avoid. In the face of this, "respectable" environmental organizations such as Friends of the Earth and Greenpeace often find themselves having to distance themselves from the activities of newer groups, such as Earth First!, TLIO, and RTS, whose lack of any organized structure makes them able and willing to break the law. The more established groups have been concerned with maintaining their scientific and political credibility, as well as with protecting their legal status and financial assets (Rootes 1997; Yearley 1996; Wall 1999). The resulting stalemates (one between the authorities and activists, one between the two sides of the environmental movement) are notable in that all the parties argue in favor of "sustainable transportation." The difference in how groups construct and draw on environmental arguments is, then, central to understanding the sustainability agenda, although this is not sufficiently recognized in much of the environmental literature (Yearley 1996; Burningham and Cooper 1999).

Policies and protests having to do with transportation and the environment deserve attention in relation to the themes of modernity and postmodernity explored in earlier chapters. A number of writers have examined the modernist basis of planning as an activity, whether in governments or in organizations more generally. Kwa (1994) writes about how, since the 1960s, planning across a range of disciplines—in the form of models of the economy, climate, transportation, and so on—has become transformed into something more postmodern. Models no longer convey grand overarching visions, focusing instead on more localized, simpler, smaller-scale approaches. In their analysis of organizational planning, Dant and Francis (1998) point to the presence of both modernist rationality and postmodern contingency within the "project" of planning—the former in terms of the rationality foregrounded in plans themselves, the latter in the ways in which they are operationalized. A significant factor in this transformation has been the decline of trust in modernist planning that was a feature of the changing culture of the 1960s and the 1970s (Healey 1977; Shackley 1997; see also papers in Campbell and Fainstein 1996).

It is perhaps not surprising, then, to find that planning policies that prioritize low-technology solutions such as the bicycle—solutions rooted in increasingly localized rather than national planning strategies—share some of their origins with environmentalist groups struggling to push such policies toward more decisive conclusions. These groups are themselves a product, along with many dimensions of contemporary cycling culture, of the search for meaning and identity described by writers such as Berger et al. (1974). Both environmental movements and transportation policy consequently involve a struggle between the modernist search for certainty and control and the postmodern flux that results from the shift away from earlier ways of organizing, protesting, and planning. That struggle is reflected in an interesting discursive tension within transportation politics that reflects wider concerns about the politics of postmodernity. How does the struggle between environmentalists and the market approach to transportation of the 1980s—"the great car economy"—map onto the shift from a modernist to a postmodernist planning paradigm? Does abandoning objectives rooted in centralized control open up land development and the provision of services such as transportation entirely for the free market, or does it instead require a new kind of control based on local constituencies and criteria that foreground issues such as sustainability and social inclusion? Obviously environmentalists would argue for the latter, and this is the direction that transportation policy has increasingly taken since the mid 1990s. How does the bicycle fit into such a model? What role does cycling have within sustainable transportation strategies? How sustainable, and how environmentally benign, is a bicycle across its full life cycle? How conducive are the various stages of this product life cycle to democracy, autonomy and community? What would a sociotechnical frame of sustainable cycling look like, and in what ways might this converge with or diverge from the frame of global flexibilization?

Sustainability and the Bicycle

Cycling is widely regarded as an inherently sustainable mode of transportation. The policy documents I have been referring to in this chapter generally point to cycling as a significant part of the overall modal shift, especially in the potential for some of the shorter trips currently made by automobile to be transferred to the bicycle—almost three-fourths of automobile trips are reported to be under 5 miles (Royal Commission on Environment and Pollution 1995: 180), a distance commonly cited as an upper limit for easy and comfortable utility cycling.

In 1996, underlining cycling's potential role in a sustainable transportation framework, the British government launched a National Cycling Strategy that featured the only national transportation-related targets yet to have been set forth: to double the rate of cycle trips by 2002 and again by 2012 (Department of Transport 1996b).[7] The strategy endorsed in the 1998 white paper followed many years of research findings in favor of cycling. In 1992 the British Medical Association had reported that cycling was an important means of gaining regular daily exercise and had argued that the health benefits of cycling outweighed the accompanying risks (BMA 1992). The Cyclists' Touring Club published several reports during the 1990s arguing the benefits of cycling not just in terms of the health of cyclists but also in terms of the savings to the National Health Service through fewer accidents and less pollution-related illness, economic savings through reduced congestion, and local and regional benefits in the improved economic performance that would result from these other factors (CTC 1991, 1993). Such findings were taken seriously in the National Cycling Strategy and in the promotion of Green Transport Plans for employers in the 1998 white paper (DETR 1998; DETR 1999).

Cycling is clearly, then, a component of the sustainability agenda, but how much is sustainability a component of the cycling agenda? How prominent are transportation, development, and the environment in the sociotechnical frame of the globally flexible bicycle? As the discussion of the relationship among mountain bikes, the environment, and wilderness shows, at first glance there seems to be a good deal of connection between cycling culture and environmental concerns. Cyclists have for many decades been concerned with the state of the remote areas they use for leisure rides, and at least the more responsible ones have engaged with local environmental issues in order to come to agreements on shared access with other kinds of users.

However, it would be difficult to argue that leisure cycling in the wilderness constitutes sustainable transportation. A central problem for activists concerned with cycling as transportation is the discrepancy between cycle ownership and cycle usage. The percentage of trips made by bicycle in the United Kingdom is very low—published data (DETR 1997; Royal Commission on Environmental Pollution 1995: 183) put it at about 2–3 percent nationally, although it is not always clear what kinds of trip such figures include, and as national figures they mask considerable locational variations (see, e.g., table 5.8b in EU 1998). Either way, there is a strong irony in the fact that in recent decades, as bicycle trips have hit a low point, bicycle ownership has risen sharply. From the early 1970s to the 1990s, the percentage of households owning a bicycle rose from 25 percent to 36 percent. Annual cycle sales globally now outstrip annual automobile sales by a factor of 3 (Bicycle Association 1997: 7).

This discrepancy between ownership and usage suggests that the sociotechnical frame of globalized flexibility in the world of cycling is somehow not promoting sustainable cycling. This is borne out by Kath Hamer's comments about Raleigh's children's bikes (chapter 5 above); it is illustrated further by the difficulty cycle designers experience in trying to establish markets for innovative designs of utility bicycles[8] (as opposed to innovative leisure bikes, which find a ready market). Most daily trips to work, to school, or to local shops are made by motor vehicle. Bicycles are at best brought out of the garage for weekend leisure rides, and at worst loaded onto the back of the car in order to be driven out to a suitably remote spot free from the traffic that such trips have themselves helped to generate. Thus, while leisure cycling is beneficial for health and allows an engagement with the countryside, it does not play any significant role in sustainable transportation, and it may even be detrimental ecologically as a result of the automobile traffic associated with it.

Activists and policy makers have been seeking ways to increase utility cycling since the 1970s. The National Cycling Strategy (NCS) can be regarded as a culmination of such efforts, which have been championed both within the United Kingdom (by the Cyclists' Touring Club, its policy offshoot the Cyclists' Public Affairs Group, and the local activist groups affiliated to the national Cycle Campaign Network) and internationally (through networks such as the annual Vélo-City conferences, which bring together activists and policy makers from around the world). The NCS was in fact developed through a good deal of consultation with cycle advocacy groups, as is evident from the range of discussion papers in its appendix (Department of Transport 1996b, 1996c). From the viewpoint of activists, the question is how to translate the policies and targets of the NCS and of the transportation white paper into a change in travel behavior away from the use of private cars and toward increased utility cycling (Davies et al. 1998)—the same problem that faces those trying to bring about modal shifts to walking, to public transportation, and to a reduced need to travel in general.

The solution to this problem that is offered by cycle activists involves promoting a wide variety of measures, including better street facilities (cycle lanes and paths, cycle racks, cycle crossings, and so on), lockers and showers in workplaces, financial incentives to encourage cycling (e.g., cycle mileage allowances comparable in value to car allowances), traffic calming, and improved public transportation to encourage motorists to use other modes. The need for such solutions is now uncontroversial (although there is still too much evidence of local highway authorities' failing to incorporate the needs of cyclists early enough in their development of new proposals). It is clear that making streets more appropriately designed and safer for cyclists, and providing facilities for cyclists at shops, leisure centers, railway stations, and workplaces will have a considerable effect on how much utility cycling takes place. Such measures are endorsed at various levels of the bicycle industry—manufacturers, retailers, and promoters of bicycles support cycling as a means of transportation for obvious commercial reasons, but they also appear to genuinely favor environmentally sound transportation as a desirable social endeavor. In the late 1990s, many manufacturers donated a small levy on each cycle sold to the National Cycle Network being built by the charity Sustrans with partial funding from the British National Lottery's Millennium Fund. More generally, the links between cycling and the environment are drawn out publicly in numerous ways by those producing, selling, promoting, and racing bicycles.

Thus, the bicycle is undoubtedly constructed—almost universally—as an inherently sustainable technology. In the broader context of transportation and mobility, however, the bicycle has only a minor role compared to that of the automobile, around which a sociotechnology involving a complex network of artifacts, infrastructures, and ways of organizing the design, the management, and the life of the city developed in the twentieth century. This sociotechnology has become so strongly embedded in Western lifestyles and culture that it is difficult to conceive any other way of structuring transportation and mobility. Cycle activists' demands for better facilities, a change of attitudes, and safer streets are problematic in this context. Cycle activists are faced, ultimately, with the question of how to counter a sociotechnology as obdurate as that of the automobile (Sørensen 1994). Despite the general consensus about the "sustainability" of bicycles, this problem is exacerbated by the ways in which bicycles are actually used in the United Kingdom—that is, primarily as leisure equipment that often must be supported by the use of an automobile. The predominance of such uses prevents cycling from making any significant contribution to a modal shift. The resulting construction of the bicycle as primarily an object for play (for both adults and children), rather than for utility, endorses and even enhances the role of the automobile in the sociotechnology of transportation and mobility.

Sustainability and the Politics of Technology

How can the problematic role of the bicycle in the wider sociotechnology of transportation and mobility be overcome? What part can the bicycle play in the development of more sustainable transportation? In view of the inconsistencies I have outlined between the assumed environmental benefits of cycling and the implications of different kinds of uses, what would actually be required of a "sustainable bicycle"? Before considering these issues, I want to return to the earlier discussion of the politics of technology, and to draw into my analysis some of the issues that have arisen along the way in tracing the history and development of bicycles and cycling.

The broad approach to technology I have taken in this book lends itself to a broad approach to sustainability, taking in not just the impacts on the natural environment of the production and use of artifacts but also their impacts on the social and cultural environments where these processes are located. In other words, how sustainable are the social structures, political relationships, and cultural values which particular technologies facilitate and reinforce? These kinds of issues come together in Bookchin's (1982) notion of *social ecology* and his call for a liberatory technology. Bookchin's account of the co-development of environmental degradation with hierarchical and hence unequal societies has parallels with eco-feminist positions (Merchant 1980) and with various green socialist perspectives (Dobson 1990). From another angle, Giddens (1990: 164) presents a vision of a "postmodern order" that combines multi-layered democratic participation, a post-scarcity economic system, demilitarization, and the humanization of technology. Such writings link the environmental aspects of sustainability to issues of equity and access in relation to technological decision making and use. Such a connection is indeed implicit in how the notion of sustainability itself was conceived: by drawing together the need to minimize ecological damage and to help the economies of the South to support themselves (World Commission on Environment and Development 1987; Yearley 1996).[9] In combining these two sets of approaches—sustainability and liberatory or democratic technology—it is not enough merely to ensure that artifacts do not damage the natural environment. The sustainability of technology is dependent on the way that each of these factors helps guarantee the other—for example, in the assumption that communities will make decisions that protect the natural environment when they themselves control the technologies that operate within it.

The complexity of the processes by which technologies become infused with cultural meaning and embedded within social and political structures is central here. Perspectives within social studies of technology can help provide an understanding of these processes. For example, Winner's (1977) notion of "epistemological Luddism" (a call for society to step back from technologies in order to consider how valuable they are) and Bijker's (1993) call to highlight both the ways in which technologies are constructed so as to favor certain groups and the idea that things could be different, are important starting points in raising critical awareness. Sclove's (1995) demands that technological decision making be made more democratic and Tatum's (1996) arguments for grassroots-based self-organization around technology and for a breaking down of the distinction between producers and consumers and that between innovators and the public go a step further by addressing the daily lives of technologies and of the social groups associated with them. Accounts of how the users of technology interact with artifacts, subvert their designated uses, further the innovation process, transform the meaning of an artifact, and deploy their own expertise in its use provide further tools for promoting the empowerment of those who do not usually contribute to crucial decisions about technology and technological change. As Feenberg (1999) suggests, paying attention simultaneously to the contingency of sociotechnical change, to ideological goals in the management of technology, and to strategies that might help achieve these goals provides the most likely route to a more democratic—and hence more sustainable—politics of technology. The various factors that must be included in such a framework are set out in table 7.1.

Toward a Sociotechnical Frame of the Sustainable Bicycle

This broader understanding of what is needed to bring about cultural change is missing from the cycle activists' wish list that I gave earlier. An ability to think through the process by which such wish lists might become embedded in systems of transportation and mobility is, however, crucial to achieving their goals. Given the above enlarging of the notion of sustainability to cover social, cultural, economic, and political as well as environmental dimensions, we might expect transportation and mobility systems to be "sustainable" in a number of ways, including the following:

- minimizing environmental damage in production and use
- maximizing access for different groups of users, and paying attention to the circumstances and concerns of non-users

Table 7.1

analytical approach

- holistic approach to technology, society, and culture
- concerned with inequalities in access and control
- concerned with everyday objects as well as large projects
- resists technological determinism
- technologies are both shaping and shaped, but there is no technological "essence"
- social and cultural embedding of technology is context specific
- technological change is contingent and emergent

sustainable-democratic technology

- is environmentally benign
- is neither socially nor environmentally damaging
- facilitates democracy, autonomy, and community
- users are empowered, not passive

practical strategies towards a sustainable-democratic politics of technology

- "epistemological Luddism"
- highlight social construction of technology and alternative possibilities
- democratize technological decision making
- grassroots self-organization
- break down barriers between innovators and users, producers and consumers
- dialogue between policy makers and affected communities

- flexible design to accommodate different users' capabilities and needs
- a sensitivity to power imbalances both within specific organizations and institutions and in the global ordering of production and consumption
- facilitating democracy, autonomy, and community among producers and users
- paying attention to these issues over the full life cycle of an artifact.

These features would have to be underpinned by an understanding of the ways in which existing systems of transportation and mobility have become socially embedded and, developing from this, an understanding of the role that bicycle technology, policy, advocacy, and activism might play in the disembedding of these systems and the re-embedding of a more sustainable and democratic transportation system. In order to bring about a sociotechnical frame of the sustainable bicycle, it is important, then, to evaluate how sustainable the current frame is, and which aspects of global flexibility are compatible or incompatible with sustainability. Then the question becomes one of how the production and the use of bicycles might be transformed into a more sustainable frame. I will address this question in relation to bicycle production and consumption and then in relation to cycling culture.

Sustainability and Production

Global flexibilization is a contradictory phenomenon from a sustainability perspective. The global structuring of the bicycle industry raises serious questions about sustainability and equity. The return of the industry (in Britain and elsewhere) from the brink of collapse since the 1980s has been achieved through the development of global networks that search the developing countries for the cheapest labor available, while the firms in control of these networks have mostly become just marketing and distribution centers. Aside from issues having to do with the politics of postcolonialism, this raises questions about the global environment, since a globally dispersed industry uses a substantial amount of energy to transport raw materials, parts, and finished goods from their place of origin to manufacturing centers and eventually to their final destination (Rosen 1995b).

Against the uneven distribution of power and control across the industry globally, it is notable that issues of power are not generally regarded as significant within the industry except in terms of labor-management relations. Although the Raleigh workforce has traditionally included a sizable proportion of women and members of ethnic minorities, none of the trade union representatives I met with were women or members of minorities. Discrimination was seen by both management and workers as not an issue at Raleigh, yet the Works Personnel Manager did concede that Raleigh had been neglecting its commitments to equal opportunity because business pressures made it difficult to divert staff time or resources in this direction. With such issues unproblematic or at least rendered invisible at Raleigh, there is more of a shared sense of destiny than before among the various industry groupings (e.g., management, labor, men, women) within the company. This does not, however, mean that workers—or even managers—feel fully in control of the technologies they use or produce. The endless adoption of new production technologies at Raleigh has allowed management to reduce the size of the company and its workforce and hence to stay profitable in the face of market changes over which both management and workers have little influence. The decision to abandon frame building intensified this situation and resulted in a less skilled workforce—a change that is bound to have a marked effect on the relations of production at the factory, not least in terms of the gender balance.

The union officials I interviewed in 1993 saw industrial relations at Raleigh as much better than the past, if not exactly harmonious. When I revisited the company in 1998, those men were gone. Their successors were less positive, having been through a wave of pay disputes on top of "redundancies" (i.e., firings intended to reduce the work force, usually on a "last in, first out" basis)—and this was before the most recent changes in the company. Management recognized that a pay dispute in the early 1990s had been poorly handled, and this seems to have been the case again in later conflicts. Nevertheless, it is increasingly understood on both sides that it is no longer in either side's interests to engage in the kinds of struggles that characterized the 1960s and the 1970s. A prolonged strike nowadays would probably force the company to close down, leaving the remaining workforce jobless. A strike has been narrowly avoided on at least two or three occasions since Raleigh's sale in 1987. What has brought the two sides together in an attempt to find less drastic solutions to labor-management disputes has been the sense of siege and uncertainty that British cycle producers feel as a result of fickle and changing markets, the success of foreign competition, and even the weather. For the remaining employees, then, there is some common interest with management.

The sense of a shared fate can be seen more strongly in a few cases of cooperative organization within the industry, and perhaps this points to one way in which sustainable industrial practices could take hold in the right circumstances. As one example, consider the existence of a network of cooperatively run cycle shops that not only display the usual traits of workers' cooperatives—shared responsibility, a lack of hierarchy among the members, a leveling of pay, and so on—but also meet informally on occasion to discuss trade, to help one another source stock items, and in some cases to produce their own ranges of bicycles. These retailers stand out among cycle shops by making a point of being hospitable to inexperienced riders and by ensuring that customers buy the right bike for their needs rather than the most fashionable one. They also play an active role in their local communities, for example supporting local racing teams. While such activities are by no means unknown among shops outside this network, they are an intrinsic feature—rather than simply a pleasant surprise—of cooperative businesses. As a second example, consider Nigel Dean Cycles, a small producer of high-quality racing and touring bikes that was turned into a workers' cooperative by the staff in the early 1990s when its founder retired. Though this producer appeared to have opted for a more formal allocation of responsibilities than the cycle shops (for example, in having demarcated roles for staff), it also attempted to set an example regarding broader relations in the industry. In 1994 it was making an effort to use components made by Sachs rather than those of Shimano.

The relatively low-key and small-scale nature of endeavors such as those just mentioned shows how much work would have to be done to focus the cycle industry on a more liberatory, equitable, and sustainable politics of technology. Such a shift would require reconfiguring both the global relations of production that have emerged since the 1970s and the local relations of production within individual firms. In order to minimize resource depletion, a globally equitable bicycle industry would also have to involve primarily small companies using local resources to supply local needs. That, in turn, would require a consumer market committed to buying locally made products rather than cheaper imports.

Sustainability and Consumption

Problems with achieving sustainability are not limited to the organization and the structuring of the cycle industry; they can also be found in relation to users. The environmental concerns that are apparently inherent to mountain biking come into question especially when this activity turns out to depend on automobile trips to "wilderness" areas—a situation that is replicated in leisure cycling on a more general scale. The interlinking of users with component design is also problematic. The user friendliness of components that has brought more people to bicycling since the mid 1980s (albeit mostly for leisure riding) has been achieved at the expense of longevity and repairability as Shimano has closed off access to innovation and design for the bicycle firms and retailers it supplies.

The globally flexible sociotechnical frame most successfully meets the sustainability criteria I have outlined in the case of catering to the heterogeneous requirements of users. Flexible production technologies mean that bicycle design can cater to a wide variety of user requirements without the extra costs this would have entailed with mass-production methods. Niche marketing within the cycle industry caters to a myriad of cycling activities and users—children, parents, the disabled, those who wish to transport unusual loads—while being sensitive to differences between men's and women's bodies (Rosen 2002b). Furthermore, the increasingly close relations between bicycle producers and consumers make it easy for users' requirements to be fed back into the processes of design and revision. There is also space in the industry for enthusiast-innovators who break down the barriers between production and consumption (ibid.), though for most consumers the products of such enterprises will be too expensive and hard to find.

Ultimately, though, it is hard to be optimistic about the globally flexible bicycle's potential to bring about sustainable transportation, since it is so closely integrated within the wider and highly embedded sociotechnology of automobile-based mobility. Despite the rising influence of sustainability discourses, it is difficult to imagine bicycles—or sustainable transportation more generally—displacing the automobile until the latter has become thoroughly disentangled from the cultural values associated with transportation and mobility. For example, social histories of the bicycle proclaim its role in liberating women in the nineteenth century (McGurn 1987), and only recently did the freedom that children gain from cycling begin to be curtailed (Hillman et al.. 1990). A sustainability-centered sociotechnical frame of the bicycle will have to wrest the values of freedom and autonomy back from the sociotechnology of the automobile and re-integrate them within a wider conception of sustainable mobility. The sheer difficulty of this task can be demonstrated by trying to imagine a car-dependent relative, friend, or colleague attaching the same sense of freedom he or she identifies with their automobile to public transportation, car sharing, or multi-modal trips. One cause for optimism is that for cyclists these values are intrinsic to their modal choice.

Sustainable Cycling Cultures

Sustainable cycling can come about only as a part of a more general shift toward sustainable mobility. The shifts in policy and public opinion—spurred on by transportation-focused direct action—that I discussed earlier in this chapter illustrate some of the ways in which moves toward sustainable mobility are being made. As these moves gain momentum in the public arena, they may stimulate the kinds of experiments with transportation technology that have taken place in California, where targets exist for the production and consumption of electric vehicles and where lanes restricted to high-occupancy automobiles are prominent (Elzen et al. 1994).[10] In Britain, road pricing (i.e., charging drivers to use specific roads) has already been established as a viable option technologically, though not yet politically (Ison 1996), and the first material result of the 1998 transportation white paper is likely to be charging for workplace parking (DETR 1998; DETR 1999). Crucial to the success of any such outcomes will be the ability to achieve a cultural shift as well as a technical or a regulatory shift. While the obduracy of the automobile is embodied materially in infrastructure, it is the culture of the automobile that secures its hold over us.

From a cultural perspective, the emergence of new meanings regarding transportation has great significance. If the automobile symbolizes modernity in terms of standardization, regulation, planning, and control in the realms of design, production, consumption, the economy, and land development, and if mountain bikes symbolize postmodernity in their celebration of flexible technology, fragmented markets, globalization, and the juxtaposition of meaning within a single artifact, what does the future hold? What kinds of cultural meanings will attach to transportation technologies in postpostmodernity? Although we cannot be sure what such a formulation will entail (Giddens 1990), transforming transportation patterns will require at least an ability to grasp the cultural dimensions of transportation and mobility, such as the meanings that different modes of transportation hold for individuals; the ways in which transportation choices are linked to identity, both individually and in wider social contexts; the contributions of particular kinds of vehicle to social status; and the ways in which transportation cultures develop collectively within organizations and across social space. This *social and cultural infrastructure* of transportation will have to be put alongside the engineering and architectural infrastructures prioritized by cycle activists as a part of a holistic analysis of the sociotechnology of transportation, or of bicycles more specifically.

The strength of cycling as a component of a possible future sociotechnology of sustainable mobility lies precisely in this cultural dimension. While the number of miles cycled has diminished, cycling culture has remained strong. Visions of bicycle futures abound among the maverick designers, competitors, and activists who populate local and national clubs focused on cycling sport, leisure, policy issues, or particular brands and models. Perhaps the strongest evidence of this culture comes from the writers, designers, and activists who have been involved in a succession of publishing ventures over the years. These include the writer and publisher Jim McGurn; Mike Burrows, the innovator who designed the bike on which the 1992 Olympic Pursuit race was won; and Richard Ballantine, the author of *Richard's Bicycle Book* (Ballantine 1983, 1988, 2000; Ballantine and Grant 1992). The work of these individuals and others was first featured in the magazine *Bicycle* in the early 1980s, then in *Cycling Today* (launched by McGurn in 1987), then in the publications of Open Road, the publishing house that once put out the magazines *Bike Culture Quarterly* and *Bycycle* and an annual catalog of unusual bikes. After Open Road ran into financial difficulties, Peter Eland, the former editor of *Bike Culture*, set up a new magazine, *Velo Vision*. Open Road's successor, the Company of Cyclists, continues to organize various events to promote cycling, including club camping holidays.

Outside the cycling fraternity, the popular association of cycling with health and the environment is a positive development. It is important not to ignore the problem that arises when this leads solely to leisure cycling; however, that can be countered where there are positive role models at hand. For example, in certain cities where levels of cycling are unusually high (e.g., Cambridge, York, and Oxford, where cycling accounts for more than 20 percent of all trips) there is an established urban cycling culture that encourages new people to take up utility cycling.

Cycle-Friendly Employers' Schemes and workplace-based Bicycle User Groups tie in with the direction being taken in government policy and might be effective even in towns with relatively few cycle commuters. It is to be hoped that the presence of a support network within an organization will provide the impetus for employees to break free of dependence on the automobile by offering (with sufficient management backing) physical and cultural infrastructures for commuting by bicycle—for example, by breaking down the way that commuting choices are generally implicated within workplace hierarchies.

It remains to be seen whether a sociotechnical frame of the bicycle will emerge that embodies the versions of sustainability and democracy I have outlined, or whether the existing frame will simply be adapted to incorporate stronger policy requirements concerning the natural environment and priority changes while retaining less equitable social and political structures. According to the model of change I have been developing in this book, the emergence of a new frame will depend on the ability of actors who are marginal to the cycling mainstream to build effectively on the kinds of alternative approaches to the production, consumption, and use of bicycles I have been discussing. Whether or not such a situation comes about, I hope to have made it clear that the technical, social, and cultural components of sociotechnical change are interdependent and that they will remain so in any reconfiguring of bicycle sociotechnology that arises out of sustainability concerns.

Conclusion: Re-Framing the Social Construction of Technology

To situate technological change within broader social, political, and cultural realms has been one of my main objectives in this book. In particular, I have tried to show how constructivist approaches can be enriched by addressing aspects of change that are usually neglected in social studies of technology. Insights from other fields and other disciplines can play a valuable role in this. In countering the questionable "purification" (Latour 1993) of technology from its social context that was built into the original exposition of the SCOT framework (Pinch and Bijker 1984), I have drawn especially on two areas of the literature.

First, I have integrated debates on modernity, postmodernity, and globalization within my analysis in order to frame the cultural, organizational, and economic "contexts" of changes that took place within the world of bicycles during the twentieth century. These contexts relate to the modernization of production, to technical and organizational innovation, to the organizational rhetorics that accompany such changes, to discursive strategies used in promoting bicycles, and to cultural changes concerning cycling, transportation, consumption, work, and identity. The "wider context" is, then, not simply a background aspect of technological change, as Pinch and Bijker seem to regard it; rather, it plays a major role in *constituting* technology. Modernity and postmodernity have been crucial to the shaping of bicycles and the various social actors concerned with them. Given the close relationship among technological change, social change, and cultural change throughout modernity,[11] technology studies as a field ought to pay far closer attention to the sociology of culture.

A second field that has strongly informed my analysis—and has indeed given the book its title—has been the bundle of work concerned with the technology and relations of production. This again is a field that has been neglected in technology studies, at least in those studies that focus their attention on discrete artifacts. Production is, though, a rich field, taking in a wide disciplinary range, and it offers much to the analysis of sociotechnical change. Studies of production offer social studies of technology ways of redressing the lack of political analysis that is characteristic of SST, where an understanding of how power structures interact with technological change is badly needed. This is something that can be found without much effort in other areas—notably, in labor process studies that question the ways in which technology is implicated in the structures of the workplace, and in gender studies that examine the gendering of technology and the role of technology in constructing gender. Literature on the technological dimensions of race and racial inequality is also becoming more common (Rosen and Skinner 2001). Where a political awareness can be detected within SST, it is usually in work that draws either on these traditions or on the strand of critical thinking about technology that can be traced back to critical theory and the counterculture. Recently political concerns have begun to surface in more mainstream SST texts, although barely any substantial politically imbued empirical work has yet been published.

SST can, in return, offer a great deal in the analysis of production, organizational change, and the politics of technology. It is especially strong in challenging simplistic accounts of causality in technological, organizational, and political change and in questioning assumptions about technological "imperatives" and the "nature" or "essence" of technology that appear to be taken for granted in other fields. (See McLaughlin et al. 1999; Grint and Woolgar 1997; Feenberg 1999.) Likewise, as I pointed out in chapter 1, the close attention in SST to the actual processes of sociotechnical change can counter the sometimes uncritical accounts of technology that are characteristic of work in the sociology of culture and (post)modernity.

What is needed within SST is further exploration of these kinds of issues, which would make it possible to build on the lessons offered by other fields and to consolidate the contribution SST can make to cultural and political analysis. As one possible element of such exploration, a number of opportunities for comparative analysis with this study immediately spring to mind.

The story of the bicycle I have told here portrays British industry adapting Fordism to its own ways of working and then being marginalized through processes of global change beyond its control. The industry's products and markets have closely matched the archetypes of standardized modernity and fragmented postmodernity. But how well do other industries, products, and consumer cultures fit these models? Although there are studies of other industries that offer a valuable basis for comparison (e.g. Smith et al. 1990), there is a great deal of scope here for further work from an SST perspective. The changing shape of transportation in recent years offers another point of comparison: How does transportation technology interact with the cultures of travel, transportation provision, and work in the transportation section, and with the politics of privatization and deregulation? And there may be other examples where sociotechnical change is set against the same kinds of relatively rapid shifts in policy or public opinion that have characterized transportation and sustainability—shifts that also result in a disjunction between established and embedded technological systems and changing cultural values. "Clean" technologies designed to meet new environmental standards are an obvious example, and these again may provide a valuable point of comparison with the bicycle case. Pursuing cases like these would enable SST researchers to build up a body of case-study analyses that engage with the cultural and political dimensions of sociotechnical change. I hope this book, which aims to account for sociotechnical change in the production and culture of a transportation technology, will soon find a place within such a body of work.

Epilogue

In the spring of 1999, Raleigh decided to stop building bicycle frames. Raleigh's Fabrication Department was closed that December, and all the production equipment was auctioned off. Attending the auction gave me a final chance to wander around a factory I had visited two or three times previously, albeit in very different circumstances. Almost exactly a year later, a second auction was held at the Sturmey-Archer factory, across the road from Raleigh. That auction told a more final story than Raleigh's. The firm that had launched the career of William Raven had become a separate entity under the ownership of Derby International. In the summer of 2000, Derby sold the company, apparently with little concern about its future survival. (For the full story, see the *Bicycle Business* Web site.) Derby had already made a substantial sum from the sale of the Sturmey site. The new buyer, Lenark (a company with few assets and a long record of buying and then liquidating firms), was unable to meet its obligations and put Sturmey into receivership in the autumn, throwing the employees out of work with ten minutes' notice. Sturmey-Archer's trademarks, its patents, its stock, and some of its equipment were then bought by the Taiwanese cycle parts manufacturer SunRace, which later renamed itself SunRace Sturmey-Archer.

The Sturmey-Archer auction included the entire contents of the factory—filing cabinets, fax machines, the telephone switchboard, computers, and canteen equipment as well as machine tools, plating and testing equipment, and conveyors. The proceeds—expected to be more than £1 million—were earmarked for creditors, with former employees high on the list. The Raleigh auction a year earlier had made only about £500,000 from the sale of manual and robotic welders, laser cutting tools, presses, drills, tube benders, washers, degreasers, and conveyor systems. Some of this equipment had been installed only a year before and was sold for barely a third of the original price. For the majority of those attending both auctions—generally people from local engineering firms rather than the cycle trade—this was a prime opportunity to get hold of high-quality engineering equipment at a discount price.

For the few who attended them out of interest in the bicycle industry, the Raleigh and Sturmey-Archer auctions were sad occasions. At the Raleigh auction, I was struck by the contrast with my previous trips to the factory, which had been a busy and noisy place—I had had to wear earplugs to avoid being deafened by the machinery. This time, the machines were silent and inactive, labeled with lot numbers. The only noise in the factory came from the auctioneer, cajoling people to bid and counterbid through a portable loudspeaker. A few works employees, standing in small groups, observed the event, but this was clearly a day off work for them. In the meantime, about 50 of their colleagues were to be made redundant by the company's change of policy. I was told that some Raleigh employees had enough work to sustain them until Christmas, while others who had worked in frame building would be transferring to other jobs—notably, inspecting the thousands of frames the company would now be importing. Many more were looking for work elsewhere—something that one man was quite sanguine about. He hadn't been made redundant himself, but he was planning to see how things worked out with the change; he had no doubts about finding other work if he decided the new-look Raleigh was not for him. How well founded his optimism was I couldn't tell.

There were some indications around the factory that most of the employees had become resigned to whatever fate awaited them. A whiteboard used to chart the progress of parts through the production process now charted the progress of equipment *out* of the factory: the words "For Sale" had been written on the board but were now struck through and replaced with "Sold . . . Gone . . . in the head." Boards detailing the responsibilities of various employees had suffered a similar fate. No doubt the workers who had been made redundant weren't happy about their situation; however, as I had found on previous visits to the factory, they seemed to accept management's claim that the situation was unavoidable—making bicycles is an uncertain business for managers and workers alike, and the alternative to redundancies in a time of fierce global competition is likely to be complete closure of the factory.

However unavoidable, Raleigh's decision to stop building frames marked the end of a tradition of more than 100 years of large-scale bicycle building in Britain. This was a turning point not just for this company but for the whole industry. Today only a few small- and medium-scale producers are building (rather than merely assembling) bikes in the United Kingdom, alongside the tiny production runs of custom frame builders. The big names who even a decade earlier were making complete bicycles had succumbed one by one to the irresistible logic of global flexibilization. Even for Raleigh, this was not a sudden jump; it was merely the end point of a process—dating back to the late 1970s (and perhaps even earlier)—that in the mid 1990s had led to a decision to end the production of front forks. If ending fork production was seen as a means of rationalizing factory-floor space while taking advantage of cheap imports, ending frame production carried this logic to its apparent conclusion. Instead of trying to compete with those whose prices will always be far cheaper than Raleigh's, it was decided to concentrate on the quality of the final product.

It is too early to consider what positive outcomes in product development or management might emerge from the changes at Raleigh. The Sturmey-Archer debacle led to revelations of serious financial difficulties at Derby International under former CEO Gary Matthews (*Bicycle Business* 16, January 2001: 4; 19, April 2001: 10–11). The treatment of Sturmey-Archer under Matthews sent shock waves throughout the cycle industry, especially at those Derby subsidiaries that remained, including Raleigh and the Dutch cycle firm Gazelle. In 2001, a management buyout at Gazelle was followed by a buyout of Derby by Alan Finden-Crofts, the group's co-founder, who had been brought in to replace Matthews.

Along with Raleigh's abandoning of production, these events indicate what an uphill task it will be to bring about any kind of shift toward a sustainable and democratic sociotechnical frame of the bicycle. British cycle firms, retailers, consumers, and activists now have little scope for encouraging sustainable domestic production or even making their own business decisions locally. Global sourcing is unavoidable for products that lie within the price range of most consumers. Opportunities for changing the dynamics of the relations of production—both nationally and globally—are thus seriously reduced. This was highlighted when one of the workers observing the Raleigh auction asked whether Raleigh was in fact going to be buying in cheaper frames from its own foreign subsidiaries and thus achieving savings for shareholders at the expense of workers both at home and abroad.

Evidence from both Raleigh and Sturmey-Archer that the cycle industry is acquiescing to and perhaps even exploiting global flexibility means that it is left to others—consumers, activists, and participants in cycling culture more broadly—to push bicycle sociotechnology toward at least a partial improvement on the globally flexible sociotechnical frame of the bicycle. With production increasingly controlled by slippery transnational corporations with agendas that have only a passing concern with sustainability, equity, or community, the central question will be whether and how bicycle advocates can ensure that this sociotechnology plays a role in the development of alternatives to car-based mobility.

Appendix A

Interviewees

Industry Commentators (1992–1998)

Strategic Industry Commentators (1992–1994, except where stated)

Serena Beeley, curator of National Cycle Museum, Lincoln

Ben Blow, Managing Director at Nigel Dean Cycles

Mike Burrows, maverick designer and innovator

Barry Forester, Managing Director at Dawes and President of Bicycle Association

Sir Frank Bowden and Gregory Houston Bowden, grandson and great-grandson of Raleigh's founder

David Collins, public relations officer at Bicycle Association of Great Britain

Lawrence Cox, Raleigh Special Products Division (1988)

Frank Ellis, Works Personnel Manager at Raleigh (1994 and 1998)

Richard Grant, cycling writer and publisher

Kath Hamer, member of York Cycleworks cooperative

Hilton Holloway, cycling writer, designer at Muddy Fox until March 1992

Anne Killick of Association of Cycle Traders

Damion McGrane, designer at Muddy Fox

Ari Hadjipetrou, Managing Director at Muddy Fox

Jim McGurn, cycling writer and publisher

Alex Moulton, independent designer and innovator

Gerald O'Donovan, Director of Raleigh's Special Products Division

Raleigh trade union representatives: convenor and two assistants from MSF (1993); after restructuring, eight Area Representatives for MSF, GMB, AEEU, TGWU (1998)

Des Reed, veteran cyclist and tinkerer with cycle technology

Yvonne Rix, Marketing Director at Raleigh (1993), interviewed again as Director of Raleigh International (1998)

Isla Rowntree, independent framebuilder and mountain bike racer

Andy Shrimpton, proprietor of Cycle Heaven shop in York

Exhibitors at the Following Cycle Shows

Cyclex, a national public show in London, March 1992

York Rally, a national rally and cycle show organized by the Cyclists' Touring Club, York, June 1992 and June 1993

Bicyclexpo, a national public show in London, October 1992

National Cycle Show (trade only) in Harrogate, organized by the Bicycle Association, June 1993

CycleFest, a festival for unusual bicycles, incorporating the British Human Power Club's annual championships, Lancaster, July 1994

Bike '95, a national public and trade show in London, April 1995

Mountain Bike Users (1992–93)

Interviews with mountain bike users helped shape my sense of pertinent issues around mountain bike culture and technology, and of how people use mountain bikes, though they were not drawn on directly in the analysis. Twenty-four respondents were recruited through a youth club, a women's cycling organization, friends, and random leafleting of bicycles at Lancaster University. In addition, I carried out a short questionnaire survey of mountain bike riders in central London.

Appendix B

Documentary Sources

Archive Sources held at Nottinghamshire County Archives and Nottingham Local Studies Library

Nottingham Local Studies Library (LSL)

Document Boxes

"Raleigh newscuttings up to 1972"

"Raleigh newscuttings 1973–1980"

"Raleigh newscuttings 1981"

"Raleigh newscuttings 1982–5"

"Raleigh newscuttings 1986–9"

"Raleigh newscuttings 1990 up to date"

"Raleigh ephemera" (contains various Raleigh catalogs, copies of the *Raligram* in-house magazine, and other documents)

"TI Annual Reports" (contains Annual Reports and statements from the 1970s and the 1980s)

Nottingham Oral History Project (OHP) Transcripts
Respondent numbers: A5, A6, A12, A15, A16, A38, A41, A43, A49, A63a, A106

Nottinghamshire County Archives
The documents and files listed in table B.1 were consulted. The full archive is cataloged at Nottinghamshire Archives; holdings up until around 1994 are also listed in Millward n.d.

Table B.1

Archive classmark	Document details	Dates[a]
DD 1267/1	List of production figures 1896-1960, plus letter about wage cuts	1931
DDRN 1/1/1	Raleigh Cycle Co. Ltd. Directors' & General Meeting Minutes	12/12/1891–18/2/1896
DDRN 1/1/2	Raleigh Cycle Co. Ltd. Directors' & General Meeting Minutes	4/3/1896–5/4/1898
DDRN 1/1/3	Raleigh Cycle Co. Ltd. Directors' & General Meeting Minutes	26/4–2/12/1898
DDRN 1/1/4	Raleigh Cycle Co. Ltd. Directors' & General Meeting Minutes	18/2/1899–12/12/1901
DDRN 1/1/5	Raleigh Cycle Co. Ltd. Directors' & General Meeting Minutes	9/1/1902–2/10/1906
DDRN 1/1/6	Raleigh Cycle Co. Ltd. Directors' & General Meeting Minutes	5/10/1906–18/8/1908
DDRN 1/2/1	Raleigh Cycle Company Minute Book No. 1	
	Fortnightly meetings	29/1/1915–1/7/1920
	Board Minutes	21/7/1919–19/3/1926
DDRN 1/2/2	RCC Minute Book No. 2	
	fortnightly minutes	1/7/1920–29/12/1922
	Board Minutes	26/3/1926–16/4/1931
DDRN 1/2/3	Raleigh Minute Book (Board Minutes)	22/4/1931–5/5/1943
DDRN 1/2/4	Raleigh Cycle Co. Board Minutes & AGMs	8/5/1943–28/6/1955
DDRN 1/2/5	Fortnightly Minutes	12/1/1922–2/3/1927
DDRN 1/2/6	Fortnightly Meetings	16/3/1927–1950s
	Board Meeting Minutes/AGMs	9/3/1953–13/1/1956
DDRN 1/3/1	Raleigh Cycle Holdings Co Ltd/ Raleigh Industries Ltd. Board Minutes	13/2/1934 – 21/12/1946
DDRN 1/9/1	British Cycle Corporation Minute Book (Board Minutes)	1956–57
DDRN 1/10/1	BSA Cycles Ltd Minute Book	1935–1972
DDRN 1/16/1	Moulton Consultants Ltd Minute Book	30/7/1957–4/10/1963

a. styled as in the archives (British style)

Table B.1
(continued)

Archive classmark	Document details	Dates[a]
DDRN 1/26/1	TI Cycle Division Council Minutes	17/1/1956–14/4/1959
DDRN 1/31/1	Raleigh Industries Ltd. Report of Directors & Statement of Accounts	1976
DDRN 1/31/2	TI Raleigh Industries Ltd. Report of Directors & Statement of Accounts	1977
DDRN 1/31/3	TI Raleigh Industries Ltd. Report of Directors & Statement of Accounts	1978
DDRN 1/31/4	TI Raleigh Industries Report of Directors & Statement of Accounts	1980
DDRN 1/32/1	TI Report to Employees, Review of . . .	1976
DDRN 1/33/1	Tube Investments Ltd. Annual Report	1979
DDRN 3/1/7	Raleigh Cycle Co. Ltd. Ledger	1923–1942
DDRN 3/1/9	Raleigh Cycle Co. Financial Ledger	1942–1952
DDRN 3/8/12	Raleigh Industries Ltd. Financial Statements & Report	31/7/1960
DDRN 3/8/13	RI Ltd. Financial Statements & Report	31/7/1961
DDRN 3/9/14	Raleigh Industries Ltd. Operating Company Accounts	1975
DDRN 3/10/14	Raleigh Industries Ltd. Non-operating Co. Accounts	1975
DDRN 3/11/1	Auditors' Reports	1926–1930
DDRN 4/2/18	Press ads	1960s
DDRN 4/2/19	Press ads	1960s
DDRN 4/17/4, 1-22, etc.	Posters & Leaflets	1970s, 1980s
DDRN 4/17/5, 1-154	Posters & leaflets	1970s, 1980s
DDRN 4/27/69-100	Posters	1970s, 1980s
DDRN 4/39/1-6	"One Free Person" TV Campaign	1975
DDRN 5/1/3	Raleigh press cuttings Vol 3.	Jan 1921–Nov 1924

Table B.1
(continued)

Archive classmark	Document details	Dates[a]
DDRN 5/1/6	Raleigh press cuttings	1931–1934
DDRN 5/1/8	Sir H Bowden press cuttings	Jan–June 1928
DDRN 5/1/9	Sir H Bowden press cuttings	July–Dec 1928
DDRN 5/3/1	Raleigh Press cuttings	4/1–31/3/1960
DDRN 5/3/2	Press cuttings	April 1960
DDRN 5/3/3	Press cuttings	May 1960
DDRN 5/3/4	Press cuttings	June 1960
DDRN 5/3/5	Press cuttings	July 1960
DDRN 5/5/11	Press Cuttings	1959
DDRN 5/5/12	Press cuttings	1960
DDRN 5/5/13	Press cuttings	1961–62
DDRN 5/5/27	Press Extracts	Jan–July 1978
DDRN 6/5/1	Album of photos as in 6/5/2	n.d. (c. 1950s)
DDRN 6/5/2	37 photos of workers & production processes	n.d. (1950s?)
DDRN 6/6/1	Photo album of opening of Triumph Road extension by Duke of Edinburgh	Nov 1952
DDRN 6/12	Numerous photographs of the Raleigh works, plus Rudge-Whitworth display boards	pre-1940
DDRN 6/28/2/1/1-117	Photographic negatives of the Raleigh factory	postwar until 1980s or 1990s
DDRN 7/3/2	Raleigh Industries Ltd Commemorative brochure, 50 Years of Leadership—Sturmey-Archer.	1952
DDRN 11/1/46	"Raleigh: A plan to fight the bosses," Labour Worker	July–Aug 1964
DDRN 11/2/13/2	Typed biography of William Raven	1940s
DDRN 11/2/18/1	Harold Bowden election candidacy card for Nottingham City Council by-election	13/6/1912

Magazines (UK, 1980s–1990s, except where stated)

Bicycle (relaunched 1993 as *Performance Cyclist International*)

Bicycle Business (launched 1999)

Bicycle Guide (US)

Bicycling (US)

Bike Culture Quarterly and *Encycleopedia* (launched 1993)

The Boneshaker—Magazine of the Veteran-Cycle Club

Cycle Industry (trade journal)

Cycle Sport

Cycle Trader (trade journal)

Cycling Plus

Cyclists' Touring Club Monthly Gazette (1887–1898) (*CTC Gazette*, 1898–1963; *Cycletouring*, 1963–1988; *Cycletouring & Campaigning*, 1988–present)

Cycling Weekly

Freewheel mail-order catalog (annual from 1978; biannual from 1992)

Mountain Bike Action (US)

Mountain Biker

Mountain Biking UK

Mountain Biking USA

MTB Pro (launched 1993)

New Cyclist (1987–1993) (*Cycling Today* 1993–present)

News & Views of the Veteran-Cycle Club

Raligram, Raleigh house magazine (1947–1962)

Which Mountain Bike? (launched 1994)

Appendix C

Significant Events and Artifacts in the History of the Raleigh Cycle Company

This chronology is based on published accounts of the company's history and on my own research material.

1886	Woodhead, Angois and Ellis set up cycle workshop in Raleigh Street, Nottingham, then move to nearby Russell Street
1888	Raleigh Cycle Company founded by partnership of Woodhead and Angois with Frank Bowden
1889	Private incorporation
1891	Undersubscribed flotation
1894	Woodhead and Angois depart
	G. P. Mills taken on as chief engineer
1896	Fully subscribed flotation
	Lenton works opened at Faraday Road
1897	Mills visits US to learn new production methods
	Launch of Gazelle brand
1898	Reconstruction of company
1900	Introduction of liquid brazing
	Introduction of sheet-steel stamping
1903	Launch of Sturmey-Archer three-speed hub
1905	Launch of Raleighette cycle-car
1906	Purchase of Robin Hood Cycles
c. 1908	William Raven employed as Works Manager
1908	Frank Bowden becomes sole owner
1914–1918	War production alongside cycle production

1915	Frank Bowden receives baronetcy
	Launch of Raleigh Light Car
	Sturmey-Archer begin producing motorcycle gears
1919	Launch of Raleigh motorcycle
1921	Death of Sir Frank Bowden
	Sir Harold Bowden takes over as Managing Director
	Sir Harold visits Henry Ford
1922	Raven leaves Raleigh
	Four-acre factory extension opened at Lenton
c. 1928–1932	Plant modernization program
	Conveyor assembly system introduced
1929	Introduction of safety brazing
	Raven returns as Works Manager
1931	Launch of three-wheel van
	New offices opened in Lenton Boulevard
1932	Humber Cycles bought out
1933	Launch of three-wheel car
1934	Company refloated as Raleigh Cycle Holdings Ltd
1935	Failed negotiations to buy Hercules Cycles
	All motorized vehicles and equipment dropped
1936	Munitions work re-introduced
1937	Launch of Sturmey-Archer Dyno-Hub
1938	Relaunch of Gazelle brand
	Sir Harold Bowden retires as Managing Director, remains Chairman
	G. H. B. Wilson takes over as Managing Director
1939–1945	Cycle production drops to 5 percent during war, with munitions production maximized
1942	William Raven retires
1943	Rudge-Whitworth Cycles bought out
	Gazelle brand renamed Robin Hood Cycles
1949–1954	Reg Harris wins World Professional Sprint Championship four times, sponsored by Raleigh

1952	New factory opened at cost of £1.25 million
1954	Cycle interests of Triumph and Three Spires bought out
1957	Cycle interests of BSA bought out
	New 20-acre factory opened at cost of £5 million
1959	Launch of Raleigh moped
	Alex Moulton's small-wheel suspension bike turned down by Raleigh
	Agreements with British Cycle Corporation over South African markets
1960	Carlton Cycles bought out
	Merger with Tube Investments
	Raleigh becomes TI's Cycle Division
	Sir Harold Bowden dies
	G. H. B. Wilson made Chairman
Early 1960s	Launch of Sunbeam range of toys
	Launch of Raleigh Roma scooter
1962	Moulton launches small-wheel suspension bicycle
1963	G. H. B. Wilson dies
1964	Seventeen-week tool-room dispute
1965	Launch of RSW 16
1966	Launch of RSW 14
1967	Moulton Cycles bought out
	Launch of Raleigh Moulton
	Launch of Raleigh 20
	Launch of Raleigh Wisp, motorized version of RSW
	Proposal of New Wages Policy
1969	Launch of Dreamline range of prams and pushchairs
	New Wages Policy imposed after rejection by workforce
1970	All motorized production dropped
	Launch of Raleigh Chopper
1974	Ilkeston R&D Factory established under Gerald O'Donovan
1977	Six-week all-out strike

1979	Loss of Nigerian and Iranian markets
	First introduction of short tracks and cellular production
1980	Office of Fair Trading report on Raleigh's sales practices
1981	Monopolies and Mergers Commission report
1985	Launch of Raleigh Maverick mountain bike
1986	R&D unit moved to Nottingham as Special Products Division
1987	Raleigh bought by Derby International
late 1980s	Adoption of robotic frame-building equipment
1989	Launch of Dyna-Tech brand
c. 1991	Introduction of adhesive bonding
1992	Launch of Raleigh Activator
1993	Launch of Activator II
1993	Launch of M-Trax range
mid 1990s	Activator range dropped
	Adhesive bonding and lugs abandoned in favor of TIG- and plasma arc welding
	Adoption of laser tube cutting and computerized toolsetting
1997	Launch of Select electrically assisted bike
	Front forks no longer made in house
1999	Raleigh stops building its own frames in favor of imported ones
2000	Derby International sells Sturmey-Archer
2001	Management buyout of Gazelle subsidiary
	Derby bought by Alan Finden-Crofts
2003	Raleigh to move to new premises after sale of Lenton factory site

Notes

Chapter 1

1. A selection of some of the most significant literature would include the following: Alderson 1972; Ballantine 1983, 1988, 1992, 2000; Beeley 1992; Bowden 1975; Bull 1991; Caunter 1955; Grew 1921; Hadland 1987; Harris and Bowden 1976; Harrison 1969, 1977, 1981, 1985; Hudson 1960; Hult 1992; Kelly and Crane 1990; Lowe 1989; Magowan 1979; McGonagle 1968; McGurn 1999; Oakley 1977; Ritchie 1975; Sanders 1991; Sharp 1977; Sillitoe 1958; Talbot 1984; Van der Plas 1988, 1991; Watson and Gray 1978; Whitt and Wilson 1974; Williamson 1966; Woodforde 1970. It is worth noting that several of the histories of the bicycle—especially those from the 1960s and the early 1970s—are regarded as inaccurate by present-day cycle historians.

2. See also Pinch and Bijker's (1984: 423) citation of this work.

3. Britain and France were the main centers of cycling activity in this period (McGurn 1999). On gender, see Bijker 1995.

4. See, e.g., DETR 1998.

Chapter 2

1. Source: *Financial Times*, April 19, 1960, held in Nottinghamshire Archives (hereafter "NA") at classmark DDRN 5/3/2. (As this note exemplifies, the British date format—with the number of the month following that of the day—is retained in citations of classmarks and similar archival data.)

2. *Nottingham Evening News*, November 19, 1962 (NA DDRN 5/5/13).

3. For a chronology of the main events in Raleigh's development, see appendix C.

4. See also NA DDRN 1/1/1, 4/4/1892.

5. Some claim that it was Michaux's employee Lallement who actually made this breakthrough (Herlihy 1995).

6. According to Sharp (1977: 157), "up till 1890 the nearest approach to the [diamond] frame was that made by Humber & Co."

7. See also NA DDRN 1/1/1, 15/1/1896.

8. NA DDRN 1/1/1, 2/10/1894, 5/1/1895, 1/7/1895, 9/8/1895.

9. *Nottingham Evening Post,* December 1, 1897 ("Raleigh newscuttings up to 1972" box, Nottingham Local Studies Library).

10. See also NA DDRN 1/1/4, 21/12/1899.

11. *Nottingham Evening Post,* December 1, 1897 ("Raleigh newscuttings up to 1972" box, Nottingham Local Studies Library).

12. NA DDRN 1/1/2, 7/10/1897.

13. Ibid.

14. NA DDRN 1/1/2, 20/7/1897, 27/7/1897, 17/8/1897, 14/9/1897.

15. *Nottingham Evening Post,* December 1, 1897, "Raleigh newscuttings up to 1972" box, Local Studies Library; NA DDRN 1/1/2, 17/9/1897.

16. NA DDRN 1/1/2, 20/7/1897.

17. NA DDRN 1/1/3, 28/10/98.

18. *Nottingham Evening Post,* December 1, 1897 ("Raleigh newscuttings up to 1972" box, Local Studies Library); NA DDRN 1/1/2, 27/1/1897.

19. NA DDRN 1/1/3, 21/4/1899. See also letter from Andrew Millward in the Veteran-Cycle Club's magazine *News and Views* 244 (December 1994–January 1995): 36–38, responding to my query about this case (ibid. 243, October-November 1994: 29–30).

20. The London daily.

21. Ibid. See also letter from Chris Watts, *News & Views* 244 (December 1994–January 1995): 35.

22. The Dutch Gazelle was, in fact, a part of the same group as Raleigh in the late twentieth century.

Chapter 3

1. *Daily Express,* October 2, 1928 (NA DDRN 5/1/9).

2. *Financial Times,* January 11, 1928 (NA DDRN 5/1/8).

3. Ibid.

4. *Nottingham Evening Post,* July 5, 1928 (NA DDRN 5/1/9).

5. NA DDRN 5/1/22.

6. NA DDRN 1/1/6, 8/1/07, 22/2/07. A typed list of Company Directors (NA DDRN 7/2/27) suggests that this occurred in 1908.

7. NA DDRN 1/2/1 (4/3/20).

8. NA DDRN 1/2/2 (1/2/22).

9. Nottinghamshire Oral History Project 1982–84, Nottingham Local Studies Library, pp. 5–6, interviewee A6.

10. NA DDRN 1/2/6 (10/10/29): 721.

11. *Nottingham Guardian,* June 30, 1938 (NA DDRN 5/1/1: 187).

12. See also NA DD 1267/21.

13. See also NA DDRN 7/3/12.

14. NA DDRN 7/2/12, p. 24.

15. Ibid., p. 40.

16. NA DDRN 1/2/4 (8/2/44).

17. *Belfast Newsletter,* December 29, 1927 (NA DDRN 5/1/8). Presumably this article was syndicated to several local papers.

18. NA DDRN 1/2/2 (9/11/21).

19. NA DDRN 1/2/5 (16/2/22): 4.

20. NA DDRN 1/2/2 (13/1/21).

21. NA DDRN 1/2/2 (13/1/21).

22. See e.g. NA DDRN 1/2/1 (21/11/19), NA DDRN 1/2/2 (16/7/20).

23. NA DDRN 1/2/6 (6/2/35): 1132.

24. Ibid. (15/3/35): 1141.

25. Ibid. (30/10/35): 1190.

26. See e.g. NA DDRN 1/2/1 (25/9/19), DDRN 1/2/2 (14/10/20).

27. NA DDRN 1/2/2 (9/3/21) (quotations from 23/3/21).

28. NA DDRN 1/2/2 (6/4/21).

29. NA DDRN 1/2/2 (31/12/20; 13/1/21; 26/1/21; 9/2/21; 23/2/21; 9/3/21; 27/7/21; 10/8/21; 14/9/21; 29/9/21; 9/11/21).

30. *Nottingham Evening News,* May 5, 1922 ("Raleigh newscuttings up to 1972" box, Nottingham Local Studies Library).

31. NA DDRN 1/2/1 (25/6/19).

32. NA DDRN 1/2/1 (4/3/20; 15/6/20); DDRN 1/2/2 (16/7/20; 10/12/20; 23/2/21).

33. NA DDRN 1/2/2 (24/11/20; 14/9/21).

34. See e.g. "The Book of the Raleigh" (1928 catalog) ("Raleigh Ephemera" box, Nottingham Local Studies Library).

35. On the Crane brothers, founders of Hercules, see Millward 1995.

36. NA DDRN 1/2/6 (28/9/27): 564.

37. *Cycling,* May 6, 1935 (NA DDRN 5/1/1: 9).

38. NA DDRN 1/2/6 (20/2/35): 1135.

39. NA DDRN 1/2/1 (18/7/19).

40. NA DDRN 1/2/5 (2/1/24): 174.

41. Ibid. (14/10/31).

42. Ibid. (7/7/32): 921.

43. NA DDRN 1/2/5 (14/9/27): 559; (28/9/27): 564; (26/10/27): 568.

44. Ibid. (29/3/28): 607.

45. NA DDRN 1/2/6 (10/10/29): 721.

46. Ibid. (10/10/29): 734.

47. See e.g. NA DDRN 1/2/1 (13/3/19), NA DDRN 1/2/2 (30/9/20).

48. NA DDRN 1/2/1 (13/2/19).

49. Ibid. (5/12/19).

50. Ibid. (1/4/20).

51. Ibid. (1/4/20; 16/4/20).

52. NA DDRN 1/2/6 (17/4/28).

53. *Walsall Pioneer*, October 21, 1921 (NA DDRN 5/1/3). In 1928, Bowden refused an unnamed American syndicate's offer of £2.5 million for his company (*Gloucestershire Echo*, November 17, 1928; *Daily News*, December 28, 1928 (NA DDRN 5/1/9).

54. NA DDRN 1/2/6 (24/2/33): 965.

55. Ibid. (16/1/34): 1033.

56. Ibid. (24/4/35): 1152–1153.

57. NA DDRN 1/2/1 (18/7/19).

58. NA DDRN 1/2/6 (25/5/27): 538.

59. Ibid. (1/5/28): 616; (16/5/28): 622; (13/6/28): 631; (13/7/28): 642.

60. Ibid. (27/6/28): 636; (30/7/28): 646.

61. NA DDRN 1/2/5 (21/7/26).

62. Ibid. (2/2/27): 503.

63. NA DDRN 1/2/6 (6/4/27): 521.

64. Ibid. (11/5/27): 534.

65. NA DDRN 1/2/1 (5/12/19; 31/12/19; 29/7/20).

66. Ibid. (18/12/19); DDRN 1/2/2 (24/11/20; 16/6/21).

67. NA DDRN 1/2/6 (5/12/28): 673.

68. Ibid. (12/10/33): 1011; (17/4/36): 1246.

69. Ibid. (31/12/25): 395.

70. NA DDRN 1/2/6 (12/7/33): 994–995; (16/1/34): 1033; (23/1/35): 1127.

71. NA DDRN 1/2/5 (12/1/22): 1.

72. NA DDRN 1/2/6 (13/7/28): 642; (28/8/28): 651; (5/12/28): 674.

73. Ibid. (2/10/30): 798.

74. Ibid. (14/10/31): 871; (21/2/32): 888.

75. Ibid. (4/9/35): 1178.

76. Raleigh Cycle Co. Ltd. 1937: vi, xi.

77. NA DDRN 1/2/6 (21/1/37).

78. *Motor Export Trader*, July 1922 (quoted in Cooper 1993: 14).

79. *Perthshire Constitutional Journal*, March 6, 1922 (NA DDRN 5/1/3: 56).

80. *Mansfield and Perch Pelts Advertiser*, November 27, 1931 (NA DDRN 5/1/6: 24).

81. *Motor Cycle and Cycle Trader*, April 3, 1932 (DDRN 5/1/6: 60).

82. *Motor Cycle and Cycle Trader*, November 10, 1933 (DDRN 5/1/6: 146).

83. See Raleigh catalogs in "Raleigh ephemera" box at Nottingham Local Studies Library.

84. These companies' catalogs can be found in NA DDRN 4/3/6 (BSA) and DDRN 4/12/11 (Hercules).

Chapter 4

1. Nottingham Local Studies Library Oral History Project (hereafter "LSL OHP") A49; A6; A5.

2. LSL OHP A5: 8.

3. LSL OHP A6: 11.

4. LSL OHP A49: 12.

5. *Labour Worker*, 1964 (NA DDRN 11/1/46).

6. LSL OHP A5: 5–6.

7. Ibid.: 2.

8. LSL OHP A63: 9–10.

9. LSL OHP A15: 7.

10. NA DDRN 1/2/5 (20/5/25): 331.

11. LSL OHP A12: 18; A41: 26.

12. LSL OHP A43: 19.

13. Ibid.

14. Ibid.: 22.

15. Ibid.

16. LSL OHP A106: 2.

17. Ibid.: 3.

18. LSL OHP A43: 23.

19. NA DDRN 1/2/2 (12/10/21).

20. Ibid. (9/11/21).

21. Ibid. (9/11/21, 14/9/21).

22. NA DDRN 1/2/5 (1/3/22): 8.

23. Ibid. (26/7/23): 142.

24. Ibid. (12/3/24): 199.

25. Ibid. (20/11/25): 382.

26. NA DDRN 1/2/6 (10/2/26): 406.

27. NA DDRN 1/2/5 (24/2/26): 411.

28. NA DDRN 1/2/6 (16/3/27): 518; (6/4/27): 523; (27/4/27): 528.

29. Ibid. (11/5/27): 534; (25/5/27): 538.

30. Ibid. (25/5/27): 540.

31. Ibid. (26/10/27): 569.

32. Ibid. (26/10/27): 569; (30/11/27): 577.

33. Ibid. (4/1/28): 586; (18/1/28): 592; (2/2/28): 596; (21/2/28): 600; (14/3/28): 603; (29/3/28): 608.

34. Ibid. (27/6/28): 639.

35. Ibid. (30/7/28): 646.

36. Ibid. (28/8/28): 651.

37. Ibid. (10/10/28): 659.

38. Ibid. (9/1/29): 676; (23/1/29): 680 ; (12/2/29): 683; (28/2/29): 687; (19/4/29): 694–695; (1/5/29): 699; (15/5/29): 802; (17/7/29): 713; (10/12/29): 732.

39. Ibid. (28/2/29): 687; (19/3/29): 691; (19/4/29): 695; (17/7/29): 713.

40. Ibid. (19/3/30): 761; (10/9/30): 793.

41. NA DDRN 1/2/2 (16/9/20).

42. NA DDRN 1/2/1 (22/1/20).

43. Ibid. (29/4/20; 14/5/20).

44. NA DDRN 1/2/5 (30/9/25): 369; (4/12/25): 387; (17/4/26): 424.

45. NA DDRN 1/2/1 (13/6/19; 18/7/19).

46. NA DDRN 1/2/5 (15/3/22): 11.

47. Ibid. (6/5/26): 429; (26/5/26): 434.

48. Administration is a mode of ordering that cannot be easily identified in data such as committee minutes, because of the nature of minutes and of administrative ordering. They are an outcome of administrative ordering, but this means their status as such is rendered invisible in the process of such ordering.

49. This ambivalence, which underpinned the shift from one sociotechnical frame to another, is illustrated by the quotations from Sir Harold Bowden in chapter 3.

Chapter 5

1. Nottingham *Guardian Journal*, May 11, 1959 (NA DDRN 5/5/11). Note that these numbers do not quite match those in table 3.2.

2. NA DDRN 1/2/6 (19/3/53).

3. Ibid. (15/9/55). The committee minutes here refer to Raleigh's competitor as TI rather than BCC. I am maintaining a distinction between the TI Cycle Division and its parent company in order not to obscure the tensions between the two that are evident from the minutes.

4. Ibid.

5. NA DDRN 1/26/1 (23/5/56, item 37; 27/6/56, item 41).

6. Ibid. (14/4/59).

7. NA DDRN 7/2/27.

8. Calculated from data in table 3.1.

9. RSW16 stood for "Raleigh small wheel 16 inches."

10. According to the 1977 Extel Statistical Services card for Raleigh, the company then owned the following cycle companies (not counting manufacturers of other products): Raleigh, Gazelle, BSA, Carlton, Moulton, New Hudson, Norman, Phillips, Rudge-Whitworth, Sunbeam, Hercules, James, Rambler, Robin Hood, Sun, Triumph, Armstrong, Dunelt.

11. *Financial Times*, March 7, 1986; TI Group plc report Progress '86: 28 (both in "Tube Investments Annual Reports" box, Nottingnam Local Studies Library).

12. *Guardian*, April 14, 1982, "Raleigh newscuttings 1982–5" box, Nottingham Local Studies Library.

13. "Main race successes: Carlton and Raleigh Professionals" leaflet (source: Gerald O'Donovan, former head of Raleigh Special Products Division).

14. *Sunday Times*, July 17, 1977; *Financial Times*, July 24, 1978 ("Raleigh newscuttings 1973–80" box); *Times*, March 14, 1986 ("Raleigh newscuttings 1986–9" box).

15. *Financial Times*, December 17, 1981 ("Raleigh newscuttings 1981" box).

16. For a summary, see MMC 1981: 36.

17. *Motor Cycle and Cycle Trader*, January 8, 1936 (NA DDRN 5/1/1: 47).

18. *Nottingham Evening Post*, December 18, 1981 ("Raleigh news cuttings 1981" box).

19. See clippings in NA DDRN 5/5/11.

20. Source: union representative quoted in *Nottingham Evening Post*, April 19, 1960 ("Raleigh up to 1972" box).

21. *Yorkshire Post*, April 8, 1961 ("Raleigh up to 1972" box).

22. Raleigh Industries Ltd. Financial Statements and Report 31/7/61: 19 (NA DDRN 3/8/13).

23. *Nottingham Evening News*, December 31, 1962 (NA DDRN 5/5/13).

24. *Guardian*, February 20, 1964 ("Raleigh up to 1972" box); Cooper 1993: 27–28.

25. See cuttings in "Raleigh up to 1972" box.

26. *Nottingham Evening Post*, December 22, 1977 ("Raleigh newscuttings 1973–80" box).

27. Tube Investments Ltd. Annual Report 1977: 13; TI Raleigh Industries Ltd. Report of Directors 1977: 4 ("TI Annual Reports" box, Nottingham Local Studies Library); NA DDRN 1/31/2.

28. TI Raleigh Industries Ltd. Report of Directors 1978: 4 (NA DDRN 1/31/3); *Guardian,* April 14, 1982; *Nottingham Evening Post,* August 31, 1983 ("Raleigh newscuttings 1982–5" box).

29. Source: interview, mid 1990s.

30. *Nottingham Evening Post,* August 31, 1983 ("Raleigh newscuttings 1982–5" box)

31. *Nottingham Voice* 72, 1977, November-December: 1 ("Raleigh newscuttings 1982–5" box); *Guardian,* October 9, 1984.

32. *Nottingham Evening Post,* August 31, 1983 ("Raleigh newscuttings 1982–5" box).

33. *Financial Times,* March 7, 1986 ("Raleigh news cuttings 1986–9" box)

34. The most recent figures here are from the *Bicycle Business* Web site.

35. TI group plc Annual Report 1986: 6 ("TI Annual Reports" box).

36. Source: 1998 interview with Frank Ellis, Works Personnel Manager.

37. *Financial Times,* March 14, 1986 ("Raleigh news cuttings 1986–9" box).

38. Source: interview with Yvonne Rix, Raleigh's Marketing Director, mid 1990s.

39. *Nottingham Evening Post,* February 26, 1988 ("Raleigh newscuttings 1986–9" box).

40. These and other views from the workforce are from interviews with representatives of trade unions.

41. See Raleigh's *Success Through Quality* newsletters of the early 1990s (NA DDRN 8/16/1–3).

Chapter 6

1. Portions of this chapter have appeared in "The social construction of mountain bikes: Technology and postmodernity in the cycle industry" (Rosen 1993).

2. Calculated using data from Garnett 1989 against those in table 6.1.

3. Calculated using date from *Cycle Press* 75 (January 1993): 6.

4. Source: *Bicycle Business* Web site.

5. Diamond Back was bought by Raleigh in 1999.

6. Source: *Bicycle Business* industry fact sheet.

7. Source: interview with Barry Forester, Managing Director of Dawes in the early 1990s.

8. Many argue that this is true of production more broadly. See e.g. Sabel and Zeitlin 1985.

9. TI Cycles of India was sold to Universal Cycles in 2001.

10. Source: letter from Hilton Holloway, designer at Muddy Fox, dated February 21, 1992.

11. This phrase was used by Hilton Holloway in an interview.

12. See Newmark 1991: 39.

13. My information about these companies came from telephone and cycle-show conversations with producers and from discussions with Roger Penn at Lancaster University.

14. This account, adapted from Rosen 1993, draws primarily on Kelly and Crane 1990.

15. From Martin 1989: 41.

16. Many of the company's innovations were in fact developed first for road use, but it is with mountain bikes that they have become most widely used.

17. Nevertheless, the timing of the decision, the night before a major cycle show, was clearly strategic.

18. For a contrasting perspective, see Winner 1986.

19. The connection between the early clunker riders and music and hippiedom is enhanced by the fact that Charles Kelly was a former roadie for rock bands.

20. NA DDRN 1/2/1 (27/2/19).

21. *Nottingham Evening Post*, July 24, 1992 ("Raleigh newscuttings since 1990" box).

22. On athletes, see Rosen 2002b.

23. TIG stands for Tungsten Inert Gas.

24. I thank Andy Shrimpton (personal communication) for pointing out the significance of BMX.

Chapter 7

1. I first came across the term "vélorution" in the UK's radical cycling press during the early 1990s. It was used by cycle activists to link cycling (i.e. the use of vélocipedes) with the direct action movement which had begun to engage with transportation issues such as government road-building programs. McGurn (1987: 178) attributes the term to the radical Montreal-based cycle activist group "La Monde à Bicyclette."

2. This rule has been wrongly presented in several accounts as requiring cyclists to ring a bell continuously. See e.g. Woodforde 1970: 3.

3. For some insight into these debates and tensions, see Redclift and Benton 1994; Yearley 1996.

4. See the text of Agenda 21 in Johnson 1993, especially chapters 4, 7, 9, and 10.

5. On these and other countercultural activities, see McKay 1996; Brass and Koziell 1997; Wall 1999.

6. Because direct action is highly dependent on the Internet for distribution of information, activist Web sites are a good source of further reading on this topic.

7. The targets have since been revised down.

8. See e.g. articles on Mike Burrows and Isla Rowntree in *New Cyclist* (17, 1991, November-December: 40–41; 27, 1992, November: 57).

9. Such multiple connections—of global capitalism, the exploitation of Third World workers, the degradation of nature and of women—are central also to Haraway's *Manifesto for Cyborgs* (1989).

10. Of course, such restrictions can be overcome in creative ways. See Latour 1992.

11. Here I adopt a broad view of modernity that accepts Giddens's view of "post-modernity" as merely an intensification of modernity.

Bibliography

Abercrombie, Nick, Scott Lash, Celia Lury, and Dan Shapiro. 1990. Flexible specialization in the culture industries? Unpublished paper, Department of Sociology, Lancaster University.

Aibar, Eduardo, and Wiebe Bijker. 1997. Constructing a city: The Cerdà plan for the extension of Barcelona. *Science, Technology, and Human Values* 22: 3–30.

Akrich, Madeleine. 1992. The de-scription of technical objects. In *Shaping Technology/Building Society*, ed. W. Bijker and J. Law. MIT Press.

Akrich, Madeleine. 1995. User representations: practices, methods, and sociology. In *Managing Technology in Society*, ed. A. Rip et al. Pinter.

Albert de la Bruheze, A., and F. Veraart. 1999. Fietsverkeer in Praktijk en Beleid in de Twintigste Eeuw. Ministerie van Verkeer en Waterstaat/Stichting Historie der Techniek, Netherlands.

Alderson, Frederick. 1972. *Bicycling: A History*. David and Charles.

Anderson, Bonnie, and Judith Zinsser. 1988. *A History of Their Own: Women in Europe from Prehistory to the Present*. Penguin.

Anderson, Håkon. 1988. Technological trajectories, cultural values, and the labour process: The development of NC machinery in the Norwegian shipbuilding industry. *Social Studies of Science* 18: 465–482.

Ashmore, Malcolm, and Evelleen Richards, eds. 1996. The politics of SSK: Neutrality, commitment, and beyond. *Social Studies of Science* 26, no. 2 (special issue).

Bailetti, Antonio Jonelle, and Paul Guild. 1991. Designers' impressions of direct contact between product designers and champions of innovation. *Journal of Product Innovation Management* 8: 91–103.

Ballantine, Richard. 1983. *Richard's Bicycle Book: A Manual of Bicycle Maintenance and Enjoyment*. Pan.

Ballantine, Richard. 1988. *Richard's New Bicycle Book*. Oxford Illustrated Press.

Ballantine, Richard. 2000. *Richard's Twenty-First Century Bicycle Book*. Pan Macmillan.

Ballantine, Richard, and Richard Grant. 1992. *Richards' Ultimate Bicycle Book*. Dorling Kindersley.

Barbrook, Richard. 1990. Mistranslations: Lipietz in London and Paris. *Science as Culture* 8: 80–117.

Baudrillard, Jean. 1988. *Selected Writings*, ed. Mark Poster. Stanford University Press.

Bauman, Zygmunt. 1991. *Modernity and Ambivalence*. Polity.

Bauman, Zygmunt. 1992. *Intimations of Postmodernity*. Routledge

Beder, Sharon. 1991. Controversy and closure: Sydney's beaches in crisis *Social Studies of Science* 21: 223–256.

Beeley, Serena. 1992. *A History of Bicycles*. Studio Editions.

Berger, Peter, Brigitte Berger, and Hansfried Kellner. 1974. *The Homeless Mind: Modernization and Consciousness*. Pelican.

Berman, Marshall. 1982. *All That Is Solid Melts into Air: The Experience of Modernity*. Verso.

Beynon, Huw. 1975. *Working for Ford*. EP.

Bicycle Association of Great Britain. 1991. Cycling: The Current Market.

Bicycle Association of Great Britain. 1997. Britain By Cycle '97.

Bijker, Wiebe. 1987. The social construction of Bakelite: Toward a theory of invention. In *The Social Construction of Technological Systems*, ed. W. Bijker et al. MIT Press.

Bijker, Wiebe. 1992. The social construction of fluorescent lighting, or how an artifact was invented in its diffusion stage. In *Shaping Technology/Building Society*, ed. W. Bijker and J. Law. MIT Press.

Bijker, Wiebe. 1993. Do not despair: There is life after constructivism. *Science, Technology, and Human Values* 18: 113–138.

Bijker, Wiebe. 1995. *Of Bicycles, Bakelites, and Bulbs: Toward a Theory of Socio-Technical Change*. MIT Press.

Bijker, Wiebe, and Karin Bijsterveld. 2000. Women walking through plans: technology, democracy, and gender identity. *Technology and Culture* 41, no. 3: 485–515.

Bijker, Wiebe, Thomas Hughes, and Trevor Pinch, eds. 1987. *The Social Construction of Technological Systems: New Directions in the Sociology and History of Technology*. MIT Press.

Bijker, Wiebe, and John Law, eds. 1992. *Shaping Technology/Building Society: Studies in Sociotechnical Change*. MIT Press.

Blume, Stuart. 1997. The rhetoric and counter-rhetoric of a "bionic" technology. *Science, Technology, and Human Values* 22, no. 1: 31–56.

BMA (British Medical Association). 1992. *Cycling Towards Health and Safety*. Oxford University Press.

Bogdanovich, Tom. 1989. Mountain bikes: Set to dominate cycling. In *The Off-Road Bicycle Book*. Leading Edge.

Bookchin, Murray. 1974. *Post-Scarcity Anarchism*. Wildwood House.

Bookchin, Murray. 1982. *The Ecology of Freedom: The Emergence and Dissolution of Hierarchy*. Cheshire Books.

Bowden, Gregory Houston. 1975. *The Story of the Raleigh Cycle.* Allen.

Bradford, Anne. 1996. *Royal Enfield: From the Bicycle to the Bullet, 1851–1969: The Story of the Company and the People Who Made It Great.* Amulree.

Brass, Elaine, and Sophie Poklewski Koziell. 1997. *Gathering Force: DIY Culture.* The Big Issue.

Braverman, Harry. 1974. *Labour and Monopoly Capital: The Degradation of Work in the Twentieth Century.* Monthly Review Press.

Bull, Andy. 1991. *Climb Every Mountain the Mountain Bike Way.* Stanley Paul.

Burawoy, Michael. 1979. *Manufacturing Consent: Changes in the Labor Process under Monopoly Capitalism.* University of Chicago Press.

Burawoy, Michael. 1985. *The Politics of Production: Factory Regimes under Capitalism and Socialism.* Verso.

Burningham, Kate, and Geoff Cooper. 1999. Being constructive: Social constructionism and the environment. *Sociology* 33, no. 2: 297–316.

Callon, Michel. 1986a. The sociology of an actor-network: The case of the electric vehicle. In *Mapping the Dynamics of Science and Technology,* ed. M. Callon et al. Macmillan.

Callon, Michel. 1986b. Some elements of a sociology of translation: domestication of the scallops and the fishermen of St. Brieuc Bay. In *Power, Action, and Belief,* ed. J. Law. Routledge and Kegan Paul.

Callon, Michel. 1987. Society in the making: The study of technology as a tool for sociological analysis. In *The Social Construction of Technological Systems,* ed. W. Bijker et al. MIT Press.

Campbell, Scott, and Susan Fainstein, eds. 1996. *Readings in Planning Theory.* Blackwell.

Caunter, C. F. 1955. *The History and Development of Cycles, As Illustrated by the Collection of Cycles in the Science Museum. Part 1: Historical Survey.* HMSO.

Clarke, Adele, and Theresa Montini. 1993. The many faces of RU486: Tales of situated knowledges and technological contestations *Science, Technology, and Human Values* 18: 42–78.

Clayton, Nick. 1999. On bicycles, Bijker, and bunkum. Paper presented to Tenth International Cycle History Conference, Nijmegen.

Cockburn, Cynthia, and Susan Ormrod. 1993. *Gender and Technology in the Making.* Sage.

Cooper, Daniel. 1993. The Growth and Decline of the Raleigh Cycle Company: A Social and Economic Analysis. Dissertation held at Nottingham Local Studies Library.

Counter Information Services. 1978. Anti-Report: The Ford Motor Company.

CTC (Cyclists' Touring Club). 1991. Bikes Not Fumes: The Emission and Health Benefits of a Modal Shift from Motor Vehicles to Cycling.

CTC. 1993. Costing the Benefits: The Value of Cycling.

Cutcliffe, Stephen. 1989. The emergence of STS as an academic field. *Research in Philosophy and Technology* 9: 287–301.

Dant, T., and D. Francis. 1998. Planning in organisations: rational control or contingent activity? *Sociological Research Online* 3, no. 2 (http://www.socresonline.org.uk/socresonline/3/2/4.html).

Davies, D., P. Emmerson, and G. Gardner. 1998. Achieving the Aims of the National Cycling Strategy: Summary of TRL Research. Report 365, Transport Research Laboratory.

Department of the Environment. 1990. This Common Inheritance: Britain's Environmental Strategy. HMSO.

Department of the Environment. 1996. Planning Policy Guidance: TownCentres and Retail Developments. HMSO.

Departments of Environment and Transport. 1994. Planning Policy Guidance: Transport. HMSO.

Department of Transport. 1989. Roads for Prosperity. HMSO.

Department of Transport. 1996a. Transport: The Way Forward. HMSO.

Department of Transport. 1996b. The National Cycling Strategy.

Department of Transport. 1996c. The National Cycling Strategy—Appendix: Topic Papers and Other Support Material.

DETR (Department of the Environment Transport and the Regions). 1997. Cycling in GB.

DETR. 1998. A New Deal For Transport: Better for Everyone. HMSO.

DETR. 1999. Preparing Your Organisation for Transport in the Future: The Benefits of Green Transport Plans.

DiMaggio, Paul, and Walter Powell. 1983. The iron cage revisited: Institutional isomorphism and collective rationality in organization fields, *American Sociological Review* 48: 147–160.

Dobson, Andrew. 1990. *Green Political Thought: An Introduction.* Unwin Hyman.

Dodge, Pryor. 1996. *The Bicycle.* Flammarion.

Du Gay, Paul, Stuart Hall, and Linda Janes. 1996. *Doing Cultural Studies: The Story of the Sony Walkman.* Sage.

Ellul, Jacques. 1964. *The Technological Society.* Vintage.

Elzen, Boelie. 1986. Two ultracentrifuges: A comparative study of the social construction of artefacts, *Social Studies of Science* 16: 621–662.

Elzen, Boelie, Johan Schot, and Remco Hoogma. 1994. Strategies for influencing the car system. In *The Car and Its Environments,* ed. K. Sørensen. EC DGXIII.

Espinoza, Zapata. 1992. Treading lightly in the land of Godzilla: Travelling to Japan to find the meaning of Shimano. *Mountain Bike Action* 7, no. 9: 32–54.

EU (European Union). 1998. Transport in Figures: Statistical Pocketbook. Office for Official Publications of the European Communities.

Eyerman, Ron, and Andrew Jamison. 1991. *Social Movements: A Cognitive Approach.* Polity.

Featherstone, Mike. 1991. *Consumer Culture and Postmodernism.* Sage.

Featherstone, Mike, Scott Lash, and Roland Robertson, eds. 1995. *Global Modernities.* Sage.

Feenberg, Andrew. 1991. *Critical Theory of Technology.* Oxford University Press.

Feenberg, Andrew. 1999. *Questioning Technology.* Routledge.

Feldberg, Roslyn, and Evelyn Nakano Glenn. 1983. Technology and work degradation: Effects of office automation on women clerical workers. In *Machina Ex Dea,* ed. J. Rothschild. Pergamon.

Ford, Henry. 1926. Mass production. *Encyclopaedia Britannica,* thirteenth edition.

Freud, Sigmund. 1901. The psychopathology of everyday life. In *The Standard Edition of the Complete Psychological Works of Sigmund Freud,* volume 6. Hogarth.

Friedman, Andrew. 1977. *Industry and Labour: Class Struggle at Work and Monopoly Capitalism.* Macmillan.

Friedman, Jonathan. 1995. Global system, globalization, and the parameters of modernity. In *Global Modernities,* ed. M. Featherstone et al. Sage.

Fujimura, Joan. 1992. Crafting science: Standardized packages, boundary objects, and "translation." In *Science as Practice and Culture,* ed. A. Pickering. University of Chicago Press

Garnett, Nick. 1989. Flat out in pursuit of the yellow jersey. *Financial Times,* October 4: 25.

Garrahan, Philip, and Paul Stewart. 1993. Working for Nissan. *Science as Culture* 16: 319–445.

Gereffi, Gary. 1994. Capitalism, development, and global commodity chains. In *Capitalism and Development,* ed. L. Sklair. Routledge.

Gibbs, Paul. 1992. Two Wheels Good? A Study of Mountain Bikes in the Natural Environment: Their Impact and Management. B.A. thesis, Leeds Metropolitan University.

Giddens, Anthony. 1990. *The Consequences of Modernity.* Polity.

Gilbert, Nigel, Roger Burrows, and Anna Pollert, eds. 1992. *Fordism and Flexibility: Divisions and Change.* Macmillan.

Glucksmann, Miriam. 1986. In a class of their own? Women workers in the new industries in inter-war Britain *Feminist Review* 24, October: 7–37.

Grew, W. F. 1921. *The Cycle Industry: Its Origins, History, and Latest Developments.* Pitman.

Grint, Keith, and Steve Woolgar. 1997. *The Machine at Work: Technology, Work, and Organization.* Polity.

Hadland, Tony. 1987. *The Sturmey-Archer Story.* Pinkerton.

Hall, Stuart, and Martin Jacques, eds. 1989. *New Times: The Changing Face of Politics in the 1990s.* Lawrence and Wishart.

Haraway, Donna. 1989. A manifesto for cyborgs: Science, technology, and socialist feminism in the 1980s. In *Coming to Terms*, ed. E. Weed. Routledge.

Harris, Reg, with Gregory Houston Bowden. 1976. *Two Wheels to the Top*. Allen.

Harris, Rosemary. 1987. *Power and Powerlessness in Industry: An Analysis of the Social Relations of Production*. Tavistock.

Harrison, Anthony. 1969. The competitiveness of the British cycle industry, 1890–1914 *Economic History Review* (second series) 22: 287–303.

Harrison, Anthony. 1977. Growth, Entrepreneurship, and Capital Formation in the United Kingdom's Cycle and Related Industries, 1870–1914. Ph.D. Thesis, University of York.

Harrison, Anthony. 1981. Joint-stock company flotation in the cycle, motor-vehicle, and related industries, 1882–1914. *Business History* 23, no. 2: 165–190.

Harrison, Anthony. 1985. The origins and growth of the UK cycle industry to 1900. *Journal of Transport History* (third series) 6, no. 1: 41–70.

Harvey, David. 1989. *The Condition of Postmodernity*. Blackwell.

Healey, Patrick. 1977. The sociology of urban transportation planning: A sociopolitical perspective. In *Urban Transport Economics*, ed. D. Hensher. Cambridge University Press.

Hemsworth, Brian. 1992. Under cover in Taiwan. *Mountain Biking USA* 6, no. 8: 62–71, 120.

Hillman, Mayer, John Adams, and John Whitelegg. 1990. *One False Move: A Study of Children's Independent Mobility*. Policy Studies Institute.

Hirst, Paul, and Grahame Thompson. 1996. *Globalization in Question: The International Economy and the Possibilities of Governance*. Polity.

Hobson, Sherren. 1992. Fiat's cultural revolution: TQM as functional integration *Science as Culture* 3, no. 14: 25–63.

Hounshell, David. 1980. The bicycle and technology in late nineteenth century America. In *Transport Technology and Social Change*, ed. P. Sörbom. Tekniska Museet.

Hounshell, David. 1984. *From the American System to Mass Production, 1800–1932: The Development of Manufacturing Technology in the United States*. Johns Hopkins University Press.

Hudson, N. B. 1960. The Growth and Structure of the Bicycle Industry. M.Sc. thesis, University College London.

Hughes, Thomas. 1986. The seamless web: Technology, science, etcetera, etcetera. *Social Studies of Science* 16: 281–292.

Hult, Jan. 1992. The Itera plastic bicycle. Social Studies of Science 22, no. 2: 373–385.

Illich, Ivan. 1973. *Tools of Conviviality*. Calder & Boyars.

Ison, Stephen. 1996. Pricing road space: Back to the future? The Cambridge experience. *Transport Reviews* 16, no. 2: 109–126.

Jameson, Fredrik. 1991. *Postmodernism, or the Cultural Logic of Late Capitalism.* Verso.

Jessop, Bob. 1991a. Fordism and post-Fordism: A critical reformulation. Working Paper 41, Lancaster Regionalism Group.

Jessop, Bob. 1991b. Thatcherism and flexibility: The white heat of a post-Fordist revolution. In *The Politics of Flexibility,* ed. B. Jessop. Elgar.

Johnson, Richard. 1986–87. What is cultural studies anyway? *Social Text* 16: 38–80.

Johnson, Stanley. 1993. *The Earth Summit: The United Nations Conference on Environment and Development.* Graham & Trotman/Martinus Nijhoff.

Joss, Simon, ed. 1999. Public Participation in Science and Technology. *Science and Public Policy* 26, no. 5 (special issue).

Kelly & Co. 1895. *Kelly's Directory of Nottinghamshire.*

Kelly, Charles, and Nick Crane. 1990. *Richard's Mountain Bike Book.* Pan.

King, Anthony. 1995. The times and spaces of modernity (or Who needs post-modernism?). In *Global Modernities,* ed. M. Featherstone et al. Sage.

Kleinman, Daniel, ed. 2000. *Science, Technology, and Democracy.* SUNY Press.

Kline, Ronald, and Trevor Pinch. 1996. Users as agents of technological change: The social construction of the automobile in the rural United States. *Technology and Culture* 37, no. 4: 763–795.

Kwa, Chunglin. 1994. Modelling technologies of control. *Science as Culture* 4, no. 20: 363–391.

Lash, Scott, and John Urry. 1987. *The End of Organized Capitalism.* Polity.

Latour, Bruno. 1987. *Science in Action: How to Follow Scientists and Engineers through Society.* Open University Press.

Latour, Bruno. 1993. *We Have Never Been Modern.* Harvester Wheatsheaf.

Latour, Bruno, and Steve Woolgar. 1979. *Laboratory Life: The Social Construction of Scientific Facts.* Sage.

Law, John. 1987. Technology and heterogeneous engineering: The case of Portuguese expansion. In *The Social Construction of Technological Systems,* ed. W. Bijker et al. MIT Press.

Law, John. 1994. *Organizing Modernity.* Blackwell.

Law, John, and Wiebe Bijker. 1992. Postscript: Technology, stability, and social theory. In *Shaping Technology/Building Society,* ed. W. Bijker and J. Law. MIT Press.

Law, John, and Michel Callon. 1992. The life and death of an aircraft: A network analysis of technical change. In *Shaping Technology/Building Society,* ed. W. Bijker and J. Law. MIT Press.

Levidow, Les, ed. 1986. *Radical Science Essays.* Free Association Books.

Lewchuk, Wayne. 1987. *American Technology and the British Vehicle Industry.* Cambridge University Press.

Lipietz, Alain. 1992. *Towards a New Economic Order: Postfordism, Ecology, and Democracy.* Polity.

Littler, Craig. 1982. *The Development of the Labour Process in Capitalist Societies.* Heinemann.

Lloyd-Jones, Roger, and M. J. Lewis. 2000. *Raleigh and the British Bicycle Industry: An Economic and Business History, 1870–1960.* Ashgate.

Lowe, Marcia. 1989. The Bicycle: Vehicle for a Small Planet. Paper 90, Worldwatch Institute.

Lupton, Tom. 1963. *On the Shop Floor: Two Studies of Workshop Organization and Output.* Pergamon.

Lutzenhiser, Loren, and Elizabeth Shove. No date. Individual Travel Behaviour: The Very Idea. Unpublished discussion paper, Washington State University and Lancaster University.

Lyotard, Jean-François. 1984. *The Postmodern Condition: A Report on Knowledge.* Manchester University Press.

MacKenzie, Donald. 1990. *Inventing Accuracy: A Historical Sociology of Nuclear Missile Guidance.* MIT Press.

MacKenzie, Donald, and Judy Wajcman, eds. 1985. *The Social Shaping of Technology: How the Refrigerator Got Its Hum.* Open University Press.

MacKenzie, Donald, and Judy Wajcman, eds. 1999. *The Social Shaping of Technology,* second edition. Open University Press.

Macnaghton, Phil, and John Urry. 1998. *Contested Natures.* Sage.

Magowan, Robin. 1979. *Tour de France: The 75th Anniversary Cycle Race.* Stanley Paul.

Mansell, Chris. 1973. The rallying of Raleigh. *Management Today,* February: 83–92.

Martin, Brian, ed. 1999. Technology and Public Participation. Science and Technology Studies, University of Wollongong.

Martin, Scott. 1989. Mountain biking turns ten. *Bicycling* 30, no. 9: 39–44.

Marx, Karl. 1970. *Capital: A Critique of Political Economy, Volume I: Capitalist Production.* Lawrence & Wishart.

Marx, Karl, and Frederick Engels. 1977. *Manifesto of the Communist Party.* Progress.

McGonagle, Seamus. 1968. *The Bicycle in Life Love War and Literature.* Pelham.

McGurn, James. 1987. *On Your Bicycle: An Illustrated History of Cycling.* John Murray.

McGurn, James. 1999. *On Your Bicycle: The Illustrated Story of Cycling,* second edition. Open Road.

McKay, George. 1996. *Senseless Acts of Beauty: Cultures of Resistance Since the Sixties.* Verso.

McLaughlin, Janice, Paul Rosen, David Skinner, and Andrew Webster. 1999. *Valuing Technology: Organisations, Culture, and Change.* Routledge.

Media Natura. 1990. The "Great Car Economy" versus The Quality of Life. Greenpeace.

Merchant, C. 1980. *The Death of Nature: Women, Ecology, and the Scientific Revolution.* Harper & Row.

Miller, Daniel. 1987. *Material Culture and Mass Consumption.* Blackwell.

Millward, Andrew. No date. The Raleigh Archives: A Detailed List of the Contents. Birmingham Polytechnic.

Millward, Andrew. 1990. The genesis of the British cycle industry 1867–1872. In Proceedings of the First International Conference of Cycling History. Museum of Transport, Glasgow.

Millward, Andrew. 1995. The founding of the Hercules Cycle and Motor Co. Ltd. In *Cycle History 5,* ed. R. van der Plas. Bicycle Books.

Misa, Thomas. 1992. Controversy and closure in technological change: Constructing "steel." In *Shaping Technology/Building Society,* ed. W. Bijker and J. Law. MIT Press.

MMC (Monopolies and Mergers Commission). 1981. Bicycles: A Report on the Application by TI Raleigh Industries Limited and TI Raleigh Limited of Certain Criteria For Determining Whether to Supply Bicycles to Retail Outlets. HMSO.

Mort, Maggie. 1994. What about the workers? Review of Graham Spinardi, *From Polaris to Trident. Social Studies of Science* 24: 596–606.

Mort, Maggie. 2002. *Building the Trident Network: The Enrollment of People, Knowledge, and Machines.* MIT Press.

Mort, Maggie, and Mike Michael. 1998. Human and technological "redundancy": Phantom intermediaries in a nuclear submarine industry. *Social Studies of Science* 28, no. 3: 355–400.

Mulkay, Michael. 1979. *Science and the Sociology of Knowledge.* Allen & Unwin.

Mumford, Lewis. 1975. Authoritarian and democratic technics. In *Technology and Culture,* ed. M. Kranzberg and W. Davenport. Meridian.

Nelis, Annemiek. 1999. Managing genetic testing: The relative powerlessness of actors in stabilised practices, *New Genetics and Society* 18, no. 2/3: 125–143.

Newmark, Liz. 1991. A dreamer with an image. *Bicycle,* midsummer: 38–40.

Noble, David. 1984. *Forces of Production: A Social History of Industrial Automation.* Knopf.

Norcliffe, Glen. 1997. Popeism and Fordism: Examining the roots of mass production. *Regional Studies* 31, no. 3: 267–280.

Nottingham Workshop. 1978. Pollution in our midst: A report by Nottingham Workshop on the pollution problems in the area of the Raleigh bicycle factory at Lenton in Nottingham.

Oakley, William. 1977. *Winged Wheel: The History of the First Hundred Years of the Cyclists' Touring Club.* Cyclists' Touring Club.

Oddy, Nicholas. 1994. Non technological factors in early cycle design. In *Cycle History: Proceedings of the Fourth International Cycle History Conference.* Bicycle Books.

Oddy, Nicholas. 1995. The bicycle: An exercise in gendered design. In *Cycle History 5,* ed. R. van der Plas. Bicycle Books.

Office of Fair Trading. 1981. A Report by the Director General of Fair Trading on an Investigation under Section 3 of the Competition Act 1980: TI Raleigh Industries Ltd./TI Raleigh Ltd. House of Commons.

Orlikowski, Wanda. 1992. The duality of technology: Rethinking the concept of technology in organizations. *Organization Science* 3, no. 3: 398–427.

Pacey, Arnold. 1999. *Meaning in Technology.* MIT Press.

Parker, Mike, and Jane Slaughter. 1990. Management-by-stress: The team concept in the US auto industry. *Science as Culture* 8: 27–58.

Patrick, Kevin. 1988. Mountain bikes and the baby boomers. *Journal of American Culture* 2, summer: 17–24.

Patton, David. 1991. A Natural History of Mountain Bikes. M.Sc. thesis, Rensselaer Polytechnic Institute.

Patton, David. 1993. Technology and tradition: Interwar cycling culture and design. Paper presented to Design History Society Conference on Moving through Design: The Culture of Transport and Travel, Southampton.

Patton, David. 1995. Aspects of a historical geography of technology: A study of *Cycling*, 1919–1939. In *Cycle History 5*, ed. R. van der Plas. Bicycle Books.

Pickering, Andrew, ed. 1992. *Science as Practice and Culture.* University of Chicago Press.

Piercy, Marge. 1979. *Woman on the Edge of Time.* Women's Press.

Piercy, Marge. 1992. *Body of Glass.* Michael Joseph.

Pinch, Trevor, and Wiebe Bijker. 1984. The social construction of facts and artefacts: or how the sociology of science and the sociology of technology might benefit each other. *Social Studies of Science* 14: 399–441.

Piore, Michael, and Charles Sabel. 1984. *The Second Industrial Divide: Possibilities for Prosperity.* Basic Books.

Raleigh Cycle Co. Ltd. 1937. Raleigh: Pioneers of Cycle Manufacture for 50 Years' Jubilee Souvenir brochure, 1887–1937.

Raleigh Industries. 1952a. The Story of a Great Enterprise. Souvenir brochure of the opening of the new factory by the Duke of Edinburgh, November 11.

Raleigh Industries. 1952b. Fifty Years of Leadership: Sturmey-Archer.

Raleigh Industries Ltd. No date (c. 1993). The art of bike building.

Rayner, Jay. 1992. Of Lycra cycling shorts and the wheels of fashion. *Independent on Sunday*, March 15, 1992: 22.

Redclift, Michael, and Ted Benton, eds. 1994. *Social Theory and the Global Environment.* Routledge.

Rip, Arie, Thomas Misa, and Johann Schot, eds. *Managing Technology in Society: The Approach of Constructive Technology Assessment.* Pinter.

Ritchie, Andrew. 1975. *King of the Road: An Illustrated History of Cycling.* Wildwood House.

Ritchie, Andrew. 1995. Developing a methodological approach to the history and meaning of velocipedes, bicycles, and tricycles. In *Cycle History 5*, ed. R. van der Plas. Bicycle Books.

Robertson, Roland. 1990. After nostalgia? Wilful nostalgia and the phases of globalization. In *Theories of Modernity and Postmodernity*, ed. B. Turner. Sage.

Rootes, Chris. 1997. Environmental movements and green parties in western and eastern Europe. In *International Handbook of Environmental Sociology*, ed. M. Redclift and G. Woodgate. Elgar.

Rosen, Paul. 1993. The social construction of mountain bikes: Technology and postmodernity in the cycle industry. *Social Studies of Science* 23: 479–513.

Rosen, Paul. 1995a. Modernity, Postmodernity, and Sociotechnical Change in the British Cycle Industry and Cycling Culture. Ph.D. thesis, Lancaster University.

Rosen, Paul. 1995b. Pedals and principles. *Bike Culture Quarterly* 5: 60–61.

Rosen, Paul. 2002a. Pro-car or anti-car? Environmental and anti-environmental discourse in UK transport debates. In *Transport Lessons for the Fuel Tax Protest of 2000*, ed. G. Lyons and K. Chatterjee. Ashgate.

Rosen, Paul. 2002b. Up the vélorution: Appropriating the bicycle and the politics of technology. In *Appropriating Technology*, ed. R. Eglash et al. University of Minnesota Press.

Rosen, Paul, and David Skinner, eds. 2001. Opening the White Box: The Politics of Racialised Science and Technology. *Science as Culture* 10, no. 3 (special issue).

Roy, Robin. 1983. Design Processes and Products Block 2: Bicycles—Invention and Innovation. Technology Coursebook for Course T263, Open University,

Roy, Robin. 1984. Product design and innovation in a mature consumer industry. In *Design Policy 2*, ed. R. Langdon. Design Council.

Royal Commission on Environmental Pollution. 1995. *Eighteenth Report: Transport and the Environment.* Oxford University Press.

Ruff, Allan, and Olivia Mellors. 1993. The mountain bike: The dream machine? *Landscape Research* 18, no. 3: 104–109.

Russell, Stewart. 1986. The social construction of artefacts: A response to Pinch and Bijker. *Social Studies of Science* 16: 331–346.

Sabel, Charles, and Jonathan Zeitlin. 1985. Historical alternatives to mass production: politics, markets, and technology in nineteenth century industrialization. *Past and Present* 108: 133–176.

Sachs, Wolfgang. 1992. *For the Love of the Automobile: Looking Back into the History of Our Desires.* University of California Press.

SACTRA. 1994. Trunk Roads and the Generation of Traffic (the SACTRA Report). HMSO.

Saetnan, Ann. 1991. Rigid politics and technological flexibility: The anatomy of a failed hospital innovation *Science, Technology, and Human Values* 16, no. 4: 419–447.

Sanders, Nick, ed. 1991. *Bicycle: The Image and the Dream.* Red Bus.

Sclove, Richard. 1995. *Democracy and Technology.* Guilford.

Seabrook, Jeremy. 2000. The migrant in the mirror. *New Internationalist* 327, September: 34–35.

Shackley, Simon. 1997. Trust in models? The mediating and transformative role of computer models in environmental discourse. In *International Handbook of Environmental Sociology*, ed. M. Redclift and G. Woodgate. Elgar.

Sharp, Archibald. 1977. *Bicycles and Tricycles: An Elementary Treatise on Their Design and Construction* (reprint of 1896 edition). MIT Press.

Sillitoe, Alan. 1958. *Saturday Night and Sunday Morning.* Signet.

Silverstone, Roger, and Eric Hirsch, eds. 1992. *Consuming Technologies: Media and Information in Domestic Space.* Routledge.

Silverstone, Roger, Eric Hirsch, and David Morley. 1992. Information and communication technologies and the moral economy of the household. In *Consuming Technologies*, ed. R. Silverstone and E. Hirsch. Routledge.

Singleton, Vicky, and Mike Michael. 1993. Actor-networks and ambivalence: General practitioners in the UK cervical screening programme. *Social Studies of Science* 23: 227–264.

Skinner, David. 1992. Technology, Consumption and the Future. Ph.D. thesis, Brunel University.

Smith, Chris, John Child, and Michael Rowlinson. 1990. *Reshaping Work: The Cadbury Experience.* Cambridge University Press.

Sørensen, Knut, ed. 1994. The Car and Its Environments: The Past, Present, and Future of the Motorcar in Europe. Proceedings of COST A4 Workshop. European Commission.

Staudenmaier, John. 1985. *Technology's Storytellers: Reweaving the Human Fabric.* MIT Press.

Street, Roger. 1998. *The Pedestrian Hobby-Horse.* Artesius.

Sutcliffe, Andy. 1991. The business machine. In *Bicycle*, ed. N. Sanders. Red Bus.

Talbot, Richard. 1984. *Designing and Building Your Own Frameset.* Manet Guild.

Tatum, Jesse. 1996. Home Power: A Model for Participatory Democracy in Technology Decision Making. Report for the Ethics and Values Study Program of National Science Foundation.

Thisdell, Dan. 1990. Shimano's dream machine. *Management Today*, March: 74–78.

Thompson, E. P. 1980. *The Making of the English Working Class.* Pelican.

Tomlinson, Alan, and Helen Walker. 1990. Holidays for all: Popular movements, collective leisure, and the pleasure industry. In *Consumption, Identity, and Style*, ed. A. Tomlinson. Routledge.

Traweek, Sharon. 1992. Border crossings: Narrative strategies in science studies and among physicists in Tsukuba Science City, Japan. In *Science as Practice and Culture*, ed. A. Pickering. University of Chicago Press.

Turner, Victor. 1957. *Schism and Continuity in an African Society: A Study of Ndembu Religious Life.* Manchester University Press.

Van der Plas, Rob. 1988. *The Mountain Bike Book*, second edition. Bicycle Books.

Van der Plas, Rob. 1991. *Bicycle Technology: Understanding, Selecting, and Maintaining the Modern Bicycle and Its Components.* Bicycle Books.

Van der Plas, Rob, ed. 1995. *Cycle History 5.* Bicycle Books.

Wajcman, Judy. 1991. *Feminism Confronts Technology.* Polity.

Wall, Derek. 1999. *Earth First! and the Anti-Roads Movement: Radical Environmentalism and Comparative Social Movements.* Routledge.

Watson, Roderick, and Martin Gray. 1978. *The Penguin Book of the Bicycle.* Penguin.

White, William, Ltd. 1894. *History, Gazetteer, and Directory of Nottinghamshire.* William White.

Whitt, Frank Rowland, and David Gordon Wilson. 1974. *Bicycling Science: Ergonomics and Mechanics.* MIT Press.

Williams, Karel, Tony Cutler, John Williams, and Colin Haslam. 1987. The end of mass production? *Economy and Society* 16, no. 3: 405–439.

Williamson, Geoffrey. 1966. *Wheels within Wheels: The Story of the Starleys of Coventry.* Geoffrey Bles.

Winner, Langdon. 1977. *Autonomous Technology: Technics-Out-of-Control as a Theme in Political Thought.* MIT Press.

Winner, Langdon. 1986. *The Whale and the Reactor: A Search for Limits in an Age of High Technology.* University of Chicago Press.

Winner, Langdon. 1993. Social constructivism: Opening the black box and finding it empty. *Science as Culture* 3, no. 16: 427–452.

Wood, Stephen, ed. 1982. *The Degradation of Work? Skill, Deskilling and the Labour Process.* Hutchinson.

Woodforde, John. 1970. *The Story of the Bicycle.* Routledge & Kegan Paul.

Woolgar, Steve. 1991. Configuring the user: The case of usability trials. In *A Sociology of Monsters,* ed. J. Law. Routledge.

World Commission on Environment and Development. 1987. *Our Common Future* (The Brundtland Report). Oxford University Press.

Wyatt, John, ed. 1992. Mountain Biking and the Environment. Report of Conference at Charlotte Mason College, Ambleside. Adventure and Environmental Awareness Group.

Wynne, Brian. 1988. Unruly technology: Practical rules, impractical discourses, and public understanding *Social Studies of Science* 18: 147–167.

Yearley, Steven. 1996. *Sociology, Environmentalism, Globalization: Reinventing the Globe.* Sage.

Zimbalist, Andrew, ed. 1979. *Case Studies on the Labor Process.* Monthly Review Press.

Index